**Electrical Principles for
Installation and Craft Studies**

Electrical Principles for Installation and Craft Studies

J. O. Paddock
T.Eng. (C.E.I.), F.I.T.E.
Department of Engineering, South Kent College of Technology

R. A. W. Galvin
B.Sc. (Hons),
Department of Engineering, South Kent College of Technology

HODDER AND STOUGHTON
London Sydney Auckland Toronto

ISBN 0 340 11518 1

First printed (as *Electrical Installation Science*) 1967
Second edition 1971
Reprinted 1973, 1975, 1976, 1978, 1980 (twice)

Printed in Great Britain
for Hodder and Stoughton Educational,
a division of Hodder and Stoughton Ltd,
Mill Road, Dunton Green, Sevenoaks, Kent
by Biddles Ltd, Guildford, Surrey

Preface

The advent of the "200 Series" of craft courses under the aegis of the CTEB, and the introduction of the SI system of units (Système International d'Unités) produced the need for a revised and extended version of the authors' previous text book, *Electrical Installation Science*. This book is the result; it provides a treatment of the 231—Electrical and Electronic Studies (Part II) syllabus for Electrical Principles and also includes the mechanical science topics called for by the Associated Studies syllabus. In addition, much of the electrical principles content of the 200—Basic Engineering Craft Studies (Part I) (Electrical Bias) syllabus has been included.

The book uses SI units throughout and the opportunity has been taken to introduce the use of the new British decimal coinage in problems involving currency. Conversion tables have been provided for converting measurements in Imperial units into their SI or metric equivalents.

The questions from examination papers used in this book have been reproduced by kind permission of the following examining bodies:

> City and Guilds of London Institutes (C.G.L.I.)
> Union of Educational Institutions (U.E.I.)
> East Midlands Educational Union (E.M.E.U.)
> Northern Counties Technical Examinations Council (N.C.T.E.C.)
> Union of Lancashire and Cheshire Institutes (U.L.C.I.)
> Welsh Joint Education Committee (W.J.E.C.)

The answers and worked solutions given for the questions included in this book are in every case those of the authors, and are not the responsibility of any examining body.

Contents

generator. Eddy currents. Characteristics of shunt, series and compound generators. Field regulators. Simple motor, practical motor. Back e.m.f., Characteristics of series, shunt and compound motors. Speed control, reversal and starting.

Units, Symbols and Abbreviations

Any measurement is essentially a process of comparison, the item being measured being compared with a standard which is itself arbitrarily defined. Although a separate standard could be defined for every type of measurement to be made this is not actually necessary; the SI system of units uses the minimum possible number of arbitrary standards, all other units being defined in terms of these standards which are:

The *metre* which is the standard of length.
The *kilogramme* which is the standard of mass.
The *second* which is the standard of time.
The *ampere* which is the standard of electric current.
The *kelvin* which is the standard of temperature.
The *candela* which is the standard of luminous intensity.

The charts below list many of the units used in electrical work, together with their symbols and abbreviations. The relationships between the derived units and the basic units from which they are derived are discussed in the appropriate chapters which follow. When using the charts it should be noted that the symbols are used to represent the quantity concerned when writing a formula or mathematical expression, for example the expression:

volts equals ohms times amperes

becomes:

$$E = RI$$

The abbreviations are used after numerical values, for example *five amperes*, becomes: 5 A.

<table>
<tr><td colspan="4" align="center">Basic (Arbitrary) Units</td></tr>
<tr><td align="center">Quantity</td><td align="center">Symbol</td><td align="center">Name of Unit</td><td align="center">Abbreviation</td></tr>
<tr><td>Length</td><td align="center">l</td><td>metre</td><td align="center">m</td></tr>
<tr><td>Mass</td><td align="center">m</td><td>kilogramme</td><td align="center">kg</td></tr>
<tr><td>Time</td><td align="center">t</td><td>second</td><td align="center">s</td></tr>
<tr><td>Electric current</td><td align="center">I</td><td>ampere</td><td align="center">A</td></tr>
<tr><td>Temperature</td><td align="center">θ</td><td>kelvin</td><td align="center">K</td></tr>
<tr><td>Luminous intensity</td><td align="center">I</td><td>candela</td><td align="center">cd</td></tr>
</table>

Derived Units			
Quantity	Symbol	Name of Unit	Abbreviation
Force	F	newton	N
Energy	W	joule	J
Power	P	watt	W
Torque	T	newton metre	N m
Quantity of electricity	Q	coulomb	C
Electric pressure	E or V	volt	V
Electrical resistance	R	ohm	Ω
Resistivity	ρ	ohm metre	Ω m
Magnetic flux	Φ	weber	Wb
Magnetic flux density	B	tesla	T
Magneto-motive-force	F	ampere	A
Magnetic field strength	H	ampere per metre	A/m
Reluctance	S	—	—
Permeability	μ	henrys per metre	H/m
Self inductance	L	henry	H
Mutual inductance	M	henry	H
Capacitance	C	farad	F
Frequency	f	hertz	Hz
Inductive reactance	X_L	ohm	Ω
Capacitive reactance	X_C	ohm	Ω
Impedance	Z	ohm	Ω
Power factor	p.f.	—	—
Specific heat capacity	c	joules per kilogramme kelvin	J/kg K
Intensity of illumination	E	lux	lx
Luminous flux	F	lumen	lm

In addition to the above units there are certain non-SI metric units which, because of their convenience are likely to be found in common use. The chart below lists the principal non-standard units with which the electrical student should be familiar.

Non-standard Units in Common Use			
Quantity	Name of unit	Abbreviation	Relation to SI unit
Mass	metric ton	t	1 metric ton = 1 000 kg
Force	kilogramme force	kgf	1 kgf = 9·81 N
Temperature	degree Celsius	°C	°C = K − 273
Quantity of heat	calorie	cal	1 cal = 4·18 J
Volume	litre	—	1 litre = 10^{-3} m³
m.m.f.	ampere turn	At	1 At = 1 A
Magnetising force	ampere turn per metre	At/m	1 At/m = 1 A/m

In many cases the unit used for measuring a particular quantity may be inconveniently large or small. In such a situation the basic unit may be either multiplied or divided by a suitable power of ten; this operation is

Prefixes for SI Units		
Prefix	Symbol	Meaning
tera	T	$\times 10^{12}$ or $\times$ 1 000 000 000 000
giga	G	$\times 10^{9}$ or $\times$ 1 000 000 000
mega	M	$\times 10^{6}$ or $\times$ 1 000 000
kilo	k	$\times 10^{3}$ or $\times$ 1 000
hecto	h	$\times 10^{2}$ or $\times$ 100
deca	da	$\times$ 10
deci	d	$\times 10^{-1}$ or $\div$ 10
centi	c	$\times 10^{-2}$ or $\div$ 100
milli	m	$\times 10^{-3}$ or $\div$ 1 000
micro	μ	$\times 10^{-6}$ or $\div$ 1 000 000
nano	n	$\times 10^{-9}$ or $\div$ 1 000 000 000
pico	p	$\times 10^{-12}$ or $\div$ 1 000 000 000 000
femto	f	$\times 10^{-15}$ or $\div$ 1 000 000 000 000 000
atto	a	$\times 10^{-18}$ or $\div$ 1 000 000 000 000 000 000

denoted by the use of prefix letters before the unit abbreviation. The chart below lists the prefixes used for this purpose together with their meanings. As the familiar Imperial units are likely to remain in common use for some time to come it is important that the student should know how to convert quantities measured in Imperial units into their SI equivalents. The chart below lists the conversion factors which are most likely to be required by the student.

Conversion Factors			
To Convert	Into	Useful Approximations	More Exact Value
lb	kg	$\times$ 0·5	$\times$ 0·4536
lbf	N	$\times$ 4·5	$\times$ 4·448
cwt	kg	$\times$ 50	$\times$ 50·80
ton (British)	kg	$\times$ 1000	$\times$ 1016
gallons	litres	$\times$ 4·5	$\times$ 4·546
gallons	m³	$\times$ 0·0045	$\times$ 0·004546
inches	cm	$\times$ 2·5	$\times$ 2·54
feet	m	$\times$ 0·3	$\times$ 0·3048
yards	m	$\times$ 0·9	$\times$ 0·9144
miles	km	$\times$ 1·6	$\times$ 1·609

1 Nature of Electricity

1.1 Conductors and Insulators

1.11 The ultimate particles from which any substance is made are called *atoms*. Some substances, called *elements*, contain only one sort of atom while others, called *compounds*, contain two or more types of atom combined together. The atoms themselves are built up of electric charges and differ only in the number of charges which they contain.

Figure 1.1 shows a simplified picture of an atom which consists of:

(a) The *nucleus* which is the central part of the atom and is made up of positive electric particles and neutral particles combined. The greater part of the mass of the atom is concentrated in the nucleus.

(b) The *electrons* which are negative electric particles that circle the nucleus in orbits.

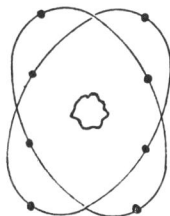

FIGURE 1.1   Structure of Atom.

1.12 The number of electrons in an atom is normally equal to the number of positive charges in the nucleus, but it is possible to remove or add electrons to the outer orbits. An atom with an *excess* of electrons is said to be *negatively* charged while an atom which is *deficient* in electrons is *positively* charged. In some materials the electrons in the outer orbits are easily moved; they tend to transfer themselves from atom to atom and so can wander freely about inside the material. Such a movement of electrons constitutes an electric current, and those materials in which electric currents can flow freely are called *conductors*. Some typical conductors are silver, copper, brass and carbon. In other materials the electrons are tightly bound to their own particular atoms, with the result that electric currents cannot flow freely in them; these materials are known as *insulators*. Some typical insulators are rubber, paper, ebonite, glass and polythene.

1.13 When electricity flows from one place to another the amount of charge transferred is equal to the number of electrons which have been moved. As the electron is such a small quantity of electricity, a unit called the **coulomb** (abbreviation C, symbol Q) is used for practical measurements. One coulomb is equal to

6.24×10^{18} electrons   (i.e. 6 240 000 000 000 000 000 electrons).

1.2 Electric Circuits

1.21 Electricity can be made to flow through a conducting wire if a force is provided which will drive the electricity into the wire at one point, and the electricity can leave the wire at some other point. The force which makes the electricity move is called an *electromotive force* (abbreviation e.m.f., symbol E).

1.22 A simple method of producing an electric current is illustrated by Fig. 1.2. The battery E provides an e.m.f. which forces the electric current through a wire into the lamp L. The electricity passes through the lamp filament making it glow and leaves via a second wire to return to the battery so completing a circuit. If either of the wires is broken or disconnected at any point the flow of electricity is interrupted and the lamp goes out.

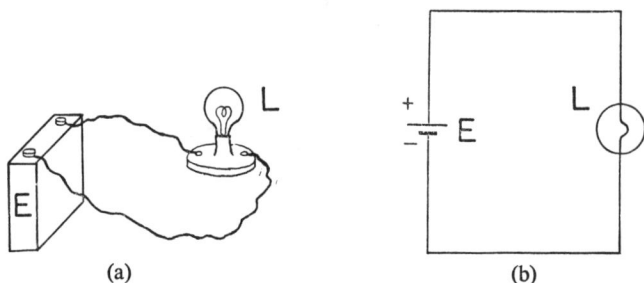

(a)   (b)
FIGURE 1.2   Simple Electrical Circuit.
(a) Practical arrangement.
(b) Theoretical diagram.

1.23 It follows that before an electric current can flow through a circuit two conditions must be fulfilled.

(a) There must be a source of e.m.f. to make the electricity move.
(b) There must be a complete path of conducting materials through which the current can flow.

1.3 Effects of an Electric Current

1.31 The effects of an electric current which are of the greatest practical use are:

(a) the heating effect,
(b) the magnetic effect,
(c) the chemical effect.

1.32 When an electric current flows through a wire heat is produced. The heating effect of a current may be demonstrated by passing a current through a lamp; the filament is heated to white heat by the passage of the current. Practical examples of the use of this effect include electric fires, cookers, water heaters, etc.

1.33 Whenever an electric current flows through a wire a magnetic field is set up around the wire. This magnetic effect can be greatly increased by forming the wire into a coil and by using an iron core. The magnetic effect can be easily demonstrated by winding a coil of about 30 turns of insulated wire around an iron rod. When current is passed through the coil the iron rod becomes magnetized and is able to attract other small pieces of iron (e.g. small nails or tacks). Practical examples of the use of the magnetic effect of a current include electro-magnets, electric motors, and electric bells.

1.34 When an electric current flows through certain liquids the liquid is separated into its chemical constituents. This process is known as *electrolysis*. The conductors which make contact with the liquid are called *electrodes*. The positive electrode by which the current enters the liquid is called the *anode*, and the negative electrode by which the current leaves the liquid is called the *cathode*. This effect may be demonstrated by passing an electric current through dilute sulphuric acid using platinum electrodes. Bubbles of hydrogen gas are released at the cathode and bubbles of oxygen are released at the anode. Practical examples of the use of electrolysis include electro-plating, preparation of chemicals, and refining of metals. Most of the aluminium produced is extracted from its ore by electro-chemical methods. This method produces an almost pure metal.

1.4 Measuring Electricity

1.41 The rate of flow of electricity through any part of a circuit is called the current and is measured in **amperes** (abbreviation A, symbol I), a current of one ampere being equivalent to a flow of one coulomb per second.

Thus:
$$I \text{ (amperes)} = \frac{Q \text{ (coulombs)}}{t \text{ (seconds)}}$$

Example 1.1

How many coulombs of electricity flow around a circuit when a current of 6 A flows for 15 seconds?

Solution

$$I = \frac{Q}{t}$$
$$\therefore \quad Q = I \times t$$
$$Q = 6 \times 15$$
$$\therefore \quad \underline{\underline{Q = 90 \text{ C}}}$$

3

1.42 In order to measure the electric current flowing through a wire it is necessary to connect an instrument called an *ammeter* in such a way that the current flows through it. Figure 1.3 shows how this can be done in the case of the simple circuit of Fig. 1.2. The current now has to flow through the ammeter on its way to the lamp, and the value of the current is indicated by the ammeter. It is important to note that the pointers of many instruments will attempt to move backwards if the direction of current flow through them is reversed, so care must be taken to connect instruments with correct polarity. The normal convention is to assume that the electric current flows from the positive (+) supply terminal to the negative (−) terminal.

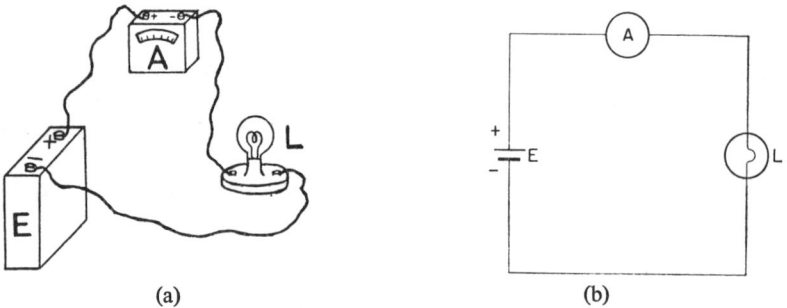

FIGURE 1.3   Measuring an Electric Current.
(a) Practical arrangement.
(b) Theoretical diagram.

1.43 Electrical pressure, such as that exerted by a source of e.m.f., is measured in **volts** (abbreviation V, symbol E or V). This pressure can be measured by using an instrument called a *voltmeter*. Figure 1.4 shows a voltmeter added to the circuit of Fig. 1.3. The voltmeter indicates the *pressure* between the battery terminals.

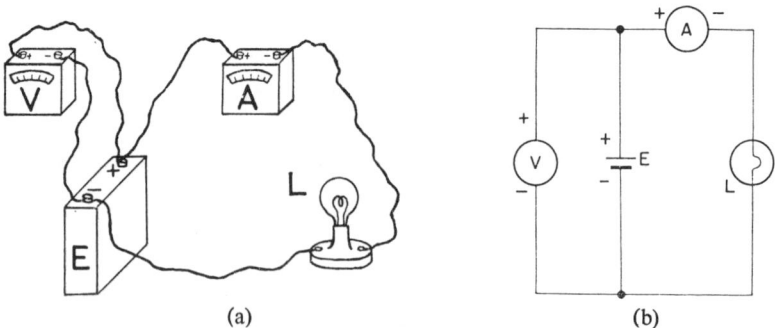

FIGURE 1.4   Measuring Electric Current and Pressure.
(a) Practical arrangement.
(b) Theoretical diagram.

1.44   Electrical power is measured in **watts** (abbreviation W, symbol P); the power produced in an electric circuit depends both on the current flowing and on the pressure.

Thus: P (watts) $= E$ (volts) $\times I$ (amperes)

The formula above shows that it is possible to determine the power in a simple circuit by measuring the electric current and pressure using an ammeter and voltmeter*; alternatively an instrument called a *wattmeter* can be used. Wattmeters have current terminals (usually marked M for mains and L for load) which are connected as an ammeter, and pressure terminals which are connected as a voltmeter. The method of connecting a wattmeter is shown in Fig. 1.5.

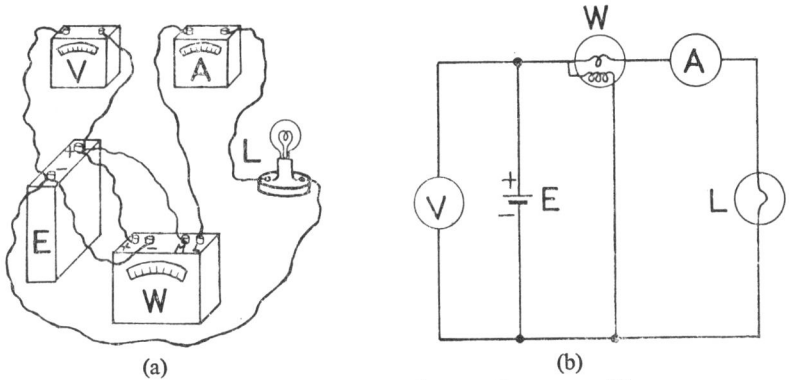

(a)   (b)

FIGURE 1.5   Measuring Electric Current, Pressure and Power.
(a) Practical arrangement.
(b) Theoretical diagram.

1.5   Resistance

1.51   Every circuit presents some opposition to the flow of electric current which has to be overcome by the electrical pressure applied to the circuit. This opposition is called the *resistance* of the circuit and is measured in **ohms,** (abbreviation Ω, symbol R).

1.52   Experiments show that the electric current flowing through a circuit is directly proportional to the electric pressure applied to it provided that the resistance is not altered in any way. This can be expressed as a formula:

$$E \text{ (volts)} = R \text{ (ohms)} \times I \text{ (amperes)}.$$

The above relationship is known as *Ohm's law* and is often expressed in words as follows:

The current (I) flowing in a closed d.c. circuit is directly proportional to the voltage (E) applied to the circuit and inversely proportional to the resistance (R) provided that the temperature does not change.

* This is strictly true only for d.c. circuits, see Chapter 8.

ELECTRICAL PRINCIPLES

This law can be expressed mathematically in three ways:

$$\text{(i)} \quad I = \frac{E}{R}$$
$$\text{(ii)} \quad E = R \times I$$
$$\text{(iii)} \quad R = \frac{E}{I}$$

the choice of formula depending on whether the value of I, E or R is required.

1.53 Electrical power is used whenever an electric current overcomes resistance, this power being converted into heat. From Section 1.44 the power is given by:

$$P = E \times I$$

But, according to Ohm's law, $E = R \times I$ and so, by substituting $R \times I$ for E in the power formula above, this becomes:

$$P = R \times I \times I$$
or
$$P = R \times I^2$$

In a similar way, by substituting $\frac{E}{R}$ for I in the power formula, this becomes:

$$P = \frac{E \times E}{R} = \frac{E^2}{R}.$$

1.54 A device specially constructed to have a definite value of resistance is called a *resistor*. Resistors are often made using lengths of resistance wire (e.g. Nichrome or Eureka wire), the wire being coiled on an insulating "former." Variable resistors or *rheostats* are used for controlling currents. A common form of rheostat is one in which bare resistance wire is wound on an insulating tube and a sliding contact makes connection with any desired point on the resistance wire. Small resistors are sometimes made using carbon or a mixture of carbon and clay as the resistance element. These resistors are usually made to carry only very small currents and are widely used in radio and electronic equipment. Some forms of resistor are illustrated in Fig. 1.6.

Example 1.2

The element of an electric fire has a resistance of 60 Ω. What value of electric current will flow through this element when it is connected to a 240 V supply and what power will it consume?

Solution

(a)
$$I = \frac{E}{R}$$
$$I = \frac{240}{60}$$
$$\therefore \quad \underline{I = 4\,A}$$

(b)
$$P = E \times I$$
$$P = 240 \times 4$$
$$\therefore \quad \underline{P = 960\,W}$$

6

Alternatively:

$$P = RI^2$$
$$P = 60 \times 4 \times 4$$
$$\therefore \quad \underline{P = 960 \text{ W}}$$

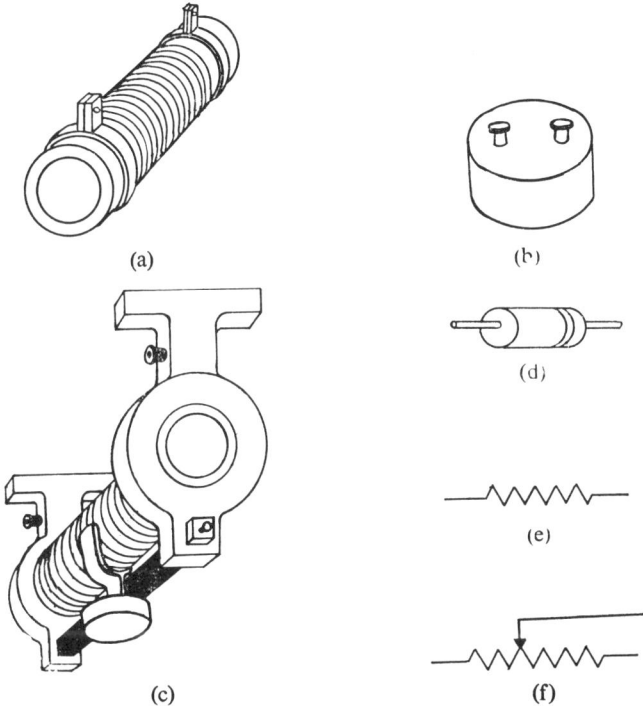

(a)

(b)

(d)

(c)

(e)

(f)

FIGURE 1.6   Resistors.

(a) Wire wound resistor.
(b) Resistance bobbin.
(c) Rheostat.
(d) Carbon composition resistor.
(e) Fixed resistor symbol.
(f) Rheostat symbol.

1.6   Electrical Energy

1.61   Electrical energy is measured in **joules** (abbreviation J, symbol W). The amount of energy produced by an electric current depends on the power being used and on the time for which it is in use.

W (work done or energy in joules) $= P$ (watts) $\times t$ (seconds)

i.e. $W = Pt.$

7

But, as $P = EI$, substituting EI for P in the above formula gives:

$$W = EIt.$$

Although the joule, or watt second, is the basic theoretical unit for the measurement of electrical energy it is too small a unit for many practical purposes. A unit called the *kilowatt hour* (abbreviation kW h) is commonly used for practical purposes, such as measuring the amount of electrical energy used by a consumer.

$$\text{kW h} = \frac{\text{watts} \times \text{hours}}{1000}$$

Example 1.3

What is the cost of operating a 3 kW immersion heater for 6 hours if electrical energy costs 0·55 p per kW h? (Answer to nearest penny.)

Solution

$$\text{kW h} = \text{kW} \times \text{h}$$

$$\text{kW h} = 3 \times 6$$

$$\underline{\text{kW h} = 18}$$

$$\therefore \quad \text{Cost} = 0·55 \times 18$$

$$\text{Cost} = 9·9 \text{ p}$$

$$\underline{\text{Cost} = 10 \text{ p (to the nearest penny)}}$$

Example 1.4

An electrical heating element having a resistance of 24 Ω is connected to a 240 V supply. Calculate:
(a) the current consumed
(b) the power consumed in kW
(c) the cost of operating the heating element for 12 h, if electrical energy costs 0·5 p per kW h (answer to nearest penny).

Solution

(a) $\quad I = \dfrac{E}{R}$

$\quad\quad I = \dfrac{240}{24}$

$\quad\quad \underline{I = 10 \text{ A}}$

(b) $P = EI$
       $P = 240 \times 10$
       $P = 2400\text{ W}$
       As $1000\text{ W} = 1\text{ kW}$, the power in kW is $2\cdot4\text{ kW}$.

(c) kW h $= 2\cdot4 \times 12$
    kW h $= 28\cdot8$
     Cost $= 0\cdot5 \times 28\cdot8$
     Cost $= 14\cdot4\text{ p}$
     Cost $= 14\text{ p}$ (to the nearest penny).

1.7 Cells

1.71 If two plates of dissimilar metals are immersed in a liquid which can react chemically with one of them, an e.m.f. is set up between the plates. Such an arrangement is called a *cell*, the plates are called *electrodes* and the liquid is called the *electrolyte*. Figure 1.7 shows a *simple cell*. The

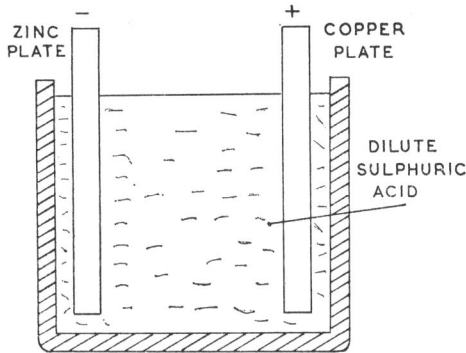

FIGURE 1.7   Simple Cell.

electrodes are a zinc plate and a copper plate and the electrolyte is dilute sulphuric acid. Such a cell produces an e.m.f. of about 1 V, but the cell has two big disadvantages which make it unsuitable for practical use. When a current is taken from the cell, bubbles of gas, in this case hydrogen, form on the copper plate, partially insulating it and also having a chemical effect which reduces the e.m.f; this effect is known as *polarisation*. Also the acid in the electrolyte tends to erode the zinc plate even though the cell is not in use; this effect is known as *local action*. The effect of local action can be reduced by coating the zinc plate with a thin layer of mercury, this process being known as *amalgamation*.

1.72 A widely used cell is the Leclanché cell. This is often made in the form of a "dry cell" as shown in Fig. 1.8. In this cell zinc is used for the negative electrode, and in the cell illustrated it also forms the casing of the

9

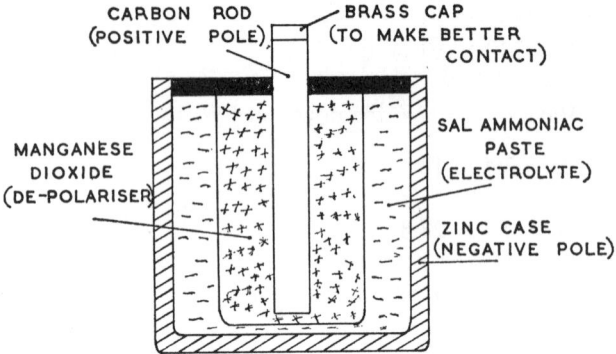

FIGURE 1.8 Dry Cell.

cell. The positive electrode is a carbon rod and the electrolyte is a paste containing sal-ammoniac. The carbon rod is surrounded by a mixture of manganese dioxide and crushed carbon. This mixture acts as a depolariser, that is it prevents polarisation by absorbing the gases which would otherwise be deposited on the carbon rod. The cell provides an e.m.f. of 1·5 V and several cells may easily be connected in series to form a *battery* giving a higher total e.m.f. where this is needed. Dry cells form a very convenient source of power for operating torches, transistor radios, etc., but are too expensive to be used for large scale power supplies.

1.73 The cells described in Sections 1.71 and 1.72 are of no further use once their chemicals are exhausted. Such cells are known as "primary"

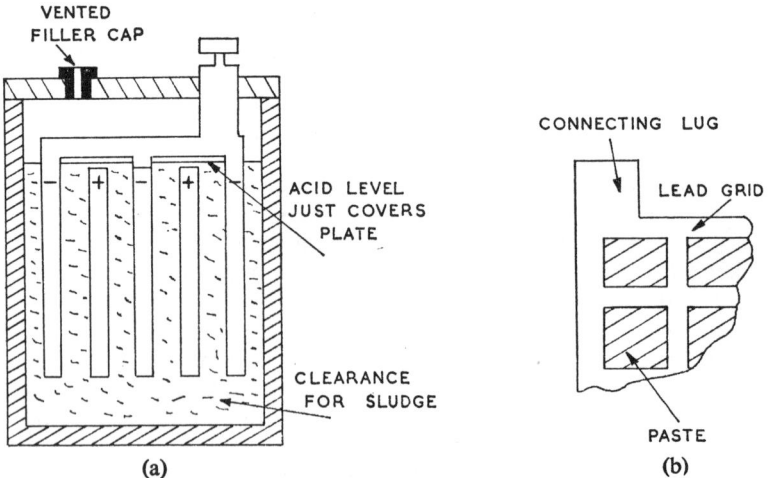

FIGURE 1.9 Lead Acid Cell.
(a) Arrangement of Plates.
(b) Detail of Plates.

cells. Storage, or "secondary", cells can be charged by passing an electric current through them in one direction and then be used to produce a discharge current in the other direction. After discharging, the cell can be recharged again ready for further use. The lead acid cell, illustrated in Fig. 1.9, consists of two sets of plates immersed in an electrolyte of dilute sulphuric acid. The plates are constructed in the form of hard lead grids which serve as frames to support the "active paste". The paste, as applied to the plates by the manufacturer, is a mixture of lead oxides and sulphuric acid. This paste sets hard and, during the initial or forming charge, is converted to lead peroxide on the positive plate and spongy lead on the negative plate.

1.74 The storage capacity of a cell depends to a large extent on the total surface area of the plates and can be increased by using a large number of plates. The e.m.f. of a freshly charged lead acid cell is approximately 2·2 V. On discharge the e.m.f. rapidly falls to about 2·0 V and remains at this figure until near the end of the useful discharge, when the e.m.f. again falls reaching 1·8 V at the end of the useful discharge. The specific gravity of the electrolyte varies during discharge from a value in the region of 1·25, that is 1·25 times as dense as water, when fully charged, to 1·10 when discharged. The specific gravity can be checked by an instrument called a hydrometer, illustrated in Fig. 1.10. The level to which the float sinks indicates the value of specific gravity. Such instruments are commonly marked in "parts per thousand" i.e. a specific gravity of 1·25

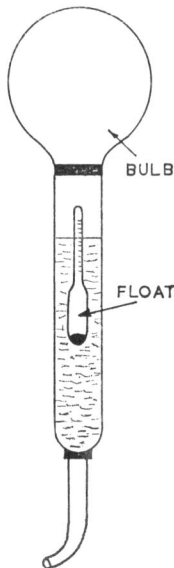

FIGURE 1.10 Hydrometer.

is indicated by a reading of 1250 and a specific gravity of 1·10 by a reading of 1100.

1.75 The table below summarises the changes which take place in a lead acid cell during charging and discharging.

Charge and Discharge of Lead Acid Cell		
Item	*Discharging*	*Charging*
Positive plates	Tend to change to lead sulphate	Change to lead peroxide
Negative plates	Tend to change to lead sulphate	Change to spongy lead
E.m.f.	Remains steady at approximately 2 V per cell until the end of the useful discharge period after which it falls rapidly. It should not be allowed to fall below 1·8 V per cell	Rises to approximately 2·7 V per cell at end of charge, but quickly falls to 2·2 V per cell when taken off charge.
Specific gravity	Falls to approximately 1·10 at the end of the useful discharge period	Rises to approximately 1·25
Characteristic curves	DISCHARGE CURVE	CHARGE CURVE

1.76 A common type of alkaline cell is the nickel cadmium cell. This cell consists of two sets of plates immersed in an electrolyte of a solution of potassium hydroxide (caustic potash) in distilled water. The plates are constructed in the form of perforated steel compartments which contain the active materials; the positive plates contain a mixture of nickel hydroxide and graphite while the negative plates contain a mixture of cadmium and iron oxides.

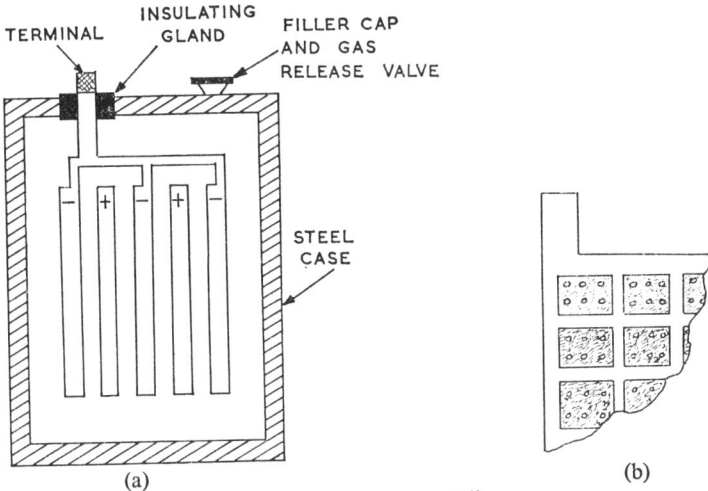

FIGURE 1.11 Alkaline Cell.
(a) Arrangement of plates.
(b) Detail of plates.

1.77 The table below summarises the changes which take place in an alkaline cell while charging and discharging.

Charge and Discharge of Alkaline Cell		
Item	*Discharging*	*Charging*
Positive plates	Lose oxygen	Restored to original condition
Negative plates	Oxidise	Restored to original condition
E.m.f.	Usually falls to approximately 1 V per cell at end of useful discharge	Rises to approximately 1·4 V per cell
Specific gravity	Remains constant at approximately 1·17	Remains constant at approximately 1·17
Characteristic curves	DISCHARGE CURVE	CHARGE CURVE

1.78 The efficiency of a secondary cell is a measure of how much can be got out of the cell on discharge compared with what was put in on charge. As the coulomb is a very small unit of charge it is often convenient to measure the amount of charge and discharge in ampere hours (abbreviation A h).

$$\text{Ampere hours of charge} = \text{charging current in amperes} \times \text{period of charge in hours}$$

The *ampere hour efficiency* of a cell is given by:

$$\text{Ampere hour efficiency} = \frac{\text{Ampere hours given on discharge}}{\text{Ampere hours required to charge}} \times 100\%$$

The ampere hour efficiency depends to some extent on the rate at which the cell is discharged, higher discharge currents giving lower efficiencies.

Example 1.5

A lead acid cell gives a discharge of 2 A for 20 h. How long will it take to recharge the cell using a current of 4 A if its ampere hour efficiency is 80%.

Solution

$$\text{Discharge} = 2 \times 20$$
$$\text{Discharge} = 40 \text{ A h}$$
$$\text{Charge required} = 40 \times \frac{100}{80}$$
$$\text{i.e. Charge required} = 50 \text{ A h}$$
$$\therefore \quad \text{Time taken} = \frac{50}{4}$$
$$\text{i.e. } t = 12{\cdot}5 \text{ h}$$

1.79 The ampere hour efficiency takes no account of the fact that it requires a higher voltage to charge the cell than is obtained on discharge. The *watt hour efficiency* takes the voltage into account and is given by:

Watt hour efficiency

$$= \frac{\text{Discharge voltage} \times \text{discharge current} \times \text{discharge time}}{\text{Charging voltage} \times \text{charging current} \times \text{time taken to recharge}} \times 100\%$$

1.710 The potential difference between the terminals of a cell on load is less than the e.m.f. of the cell because of the effect of *internal resistance*. The current must flow through the resistance of the internal parts of the cell as well as through the external circuit, and some of the cell's e.m.f. will be used to overcome this internal resistance. By Ohm's law, the pressure used to overcome the internal resistance, r, is $V = rI$, thus the p.d. remaining across the cell terminals is:

$$V = E - rI.$$

Example 1.6

Describe a lead acid secondary cell.

Explain briefly the changes in the cell during charge and discharge.

The potential difference at the terminals of a lead-acid cell on open circuit was 2·18 V; when the cell was discharging at the rate of 9 A the terminal p.d. was 2·02 V. Calculate the internal resistance of the cell.

(C.G.L.I.)

Solution

The descriptive parts of the question are already covered in Section 1.73 and will not be repeated here.

The solution to the numerical part of the question is as follows:

$$V = E - rI$$

When no current is taken, the p.d. between the cell terminals is equal to E, as with no current flowing there can be no loss of voltage in the internal resistance. Substituting the values given into the equation gives:

$$2·02 = 2·18 - 9r$$
$$9r = 2·18 - 2·02$$
$$9r = 0·16$$
$$\therefore \quad r = 0·0178 \text{ ohm}$$

1.8 Electrolysis

1.81 As has been already stated in Section 1.34, electrolysis means the splitting up of a chemical solution into its constituents by passing an electric current through it. Faraday's laws of electrolysis can be used to determine the amount of substance liberated by electrolysis. They are:

(a) The mass of substance liberated at the cathode is proportional to the quantity of electricity passed through the electrolyte.

(b) The mass of substance liberated at the cathode is proportional to the electrochemical equivalent mass of the substance.

The electrochemical equivalent mass (abbreviation E.C.E., symbol z) is the mass of substance liberated by the passage of one coulomb of electricity through the electrolyte.

1.82 Figure 1.12 shows a simple method of determining the E.C.E. of copper. The cathode is weighed before the current is switched on, and again after the current has been passed for a known time, the mass of copper deposited being determined by the gain in mass. The quantity of electricity passed in coulombs is calculated using:

$$Q = It$$

Thus:
$$z = \frac{\text{Mass deposited, in grammes}}{\text{coulombs passed through electrolyte}}$$

15

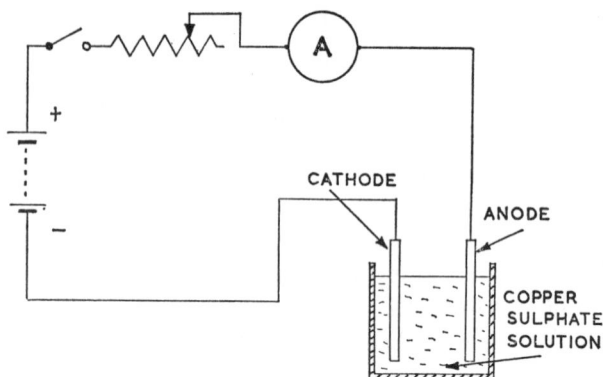

BOTH ANODE AND CATHODE ARE COPPER PLATES
FIGURE 1.12 Determination of E.C.E. of Copper.

For this experiment to be successful the following precautions must be observed:

(a) The electrodes must be thoroughly cleaned before starting the experiment.
(b) The cathode must be carefully washed and dried before the final weighing, care being taken not to disturb any of the deposit in the process. The cathode plate should first be thoroughly rinsed in water, rapid drying can then be promoted by rinsing in methylated spirits. On no account should the plate be heated in an attempt to speed up the drying as this may cause the deposit to oxidise.
(c) The current must be kept low, to ensure a firm deposit.
(d) The time for which the current is allowed to flow should be as long as possible.
(e) As the mass of deposit is found as the difference between two much larger quantities, the 'weighing' must be carried out as accurately as possible using a chemical balance.

1.83 Faraday's laws of electrolysis may be expressed as a single equation:

$$m = Itz$$

It is usual to express z in grammes per coulomb, and in this case m is the mass deposited on the cathode in grammes.

Example 1.7

A current of 2 A is passed through a copper sulphate solution for 3 hours, what mass of copper is deposited on the cathode? E.C.E. of copper = 0·000 33 grammes per coulomb.

Solution

$$m = Itz$$
$$m = 2 \times 3 \times 3600 \times 0·000\ 33$$

Note that "t" is expressed in seconds

$$m = 7\cdot13 \text{ g}$$

SUMMARY OF FORMULAE

Quantity of electricity: $\qquad Q = It$ coulombs

Electrical power.
$$\begin{cases} P = EI \text{ watts} \\ P = RI^2 \text{ watts} \\ P = \dfrac{E^2}{R} \text{ watts} \\ kW = \dfrac{\text{watts}}{1000} \end{cases}$$

Ohm's law.
$$\begin{cases} I = \dfrac{E}{R} \text{ amperes} \\ E = RI \text{ volts} \\ R = \dfrac{E}{I} \text{ ohms} \end{cases}$$

Electrical energy.
$$\begin{cases} W = Pt \text{ joules} \\ kW h = \dfrac{\text{watts} \times \text{hours}}{1000} \end{cases}$$

Effect of internal resistance. $\qquad V = E - rI$

Electrolysis. $\qquad m = Itz$

EXERCISE 1

1. Complete the following statements:
   (a) A material through which electricity flows freely is called a ....
   (b) A material through which electricity cannot flow freely is called an ....
   (c) The ultimately smallest particle of a substance is an ....
   (d) The central part of the atom is called the .... and the .... move around this in orbits.
   (e) The electron has a .... electric charge.
   (f) The nucleus has a .... electric charge.
2. What conditions must be fulfilled before an electric current can flow through a circuit?
3. What are the three main effects of an electric current? Give one practical application of each of the above effects.
4. Complete the following statements:
   (a) Quantity of electricity is measured in ....
   (b) Electric current is measured in ....
   (c) Electro-motive-force is measured in ....
   (d) Electrical resistance is measured in ....
   (e) Electrical power is measured in ....
   (f) Electrical energy is measured in ....

5. State Ohm's law in words and in symbols. Calculate the missing values in the table below, and enter them in the appropriate space.

E	6 V	15 V		240 V		150 V	
I	3 A		5 A		8 A		
R			12 Ω	100 Ω			25 Ω
P		45 W			128 W	75 W	100 W

6. Show, by means of a theoretical circuit diagram and a sketch of the practical arrangement, how to connect a battery to a resistance bobbin together with the necessary instruments to measure the electric current and pressure. How could the resistance of the resistor be calculated from the meter readings?

7. An electric heater is rated at 1 kW, 240 V. What current is this heater intended to consume and what is the resistance of the heating element. If the heater is connected to a 200 V supply, its resistance being unchanged, what power would it now consume?

8. A 3 kW immersion heater is used for 14 hours per week for a period of 13 weeks. What is the cost if the price of electrical energy is 0·52 p per kW h?

9. In an electro-plating plant, a current of 200 amperes flows for 1½ hours. Calculate the quantity of electricity (a) in ampere hours, (b) in coulombs. If the voltage applied is 30, calculate the energy in kilowatt hours.   N.C.T.E.C.

10. Sketch one each of any form of:
    (a) Primary cell.
    (b) Secondary cell.
    Clearly label the component parts. Describe how each cell operates and the value of the e.m.f. in each case.   C.G.L.I

11. Describe the construction of one form of secondary cell. Explain the factors governing its capacity.
    A battery is rated at 200 A h with discharge current 10 A. Determine the number of hours it will discharge at the 10 A rate. Explain what you would expect to happen if the current is doubled to 20 A.
    Explain the term "ampere hour efficiency".   E.M.E.U.

12. (a) What is meant by *electrolysis*?
    (b) Describe with a labelled diagram the construction of a dry Leclanché cell.
                          N.C.T.E.C.

2 Electric Circuits I

2.1. Series Circuits

2.11 In a series circuit the connections are such that the electric current flows through each part of the circuit in turn. Figure 2.1 shows three resistors R_1, R_2 and R_3 connected in series with a source of e.m.f. E

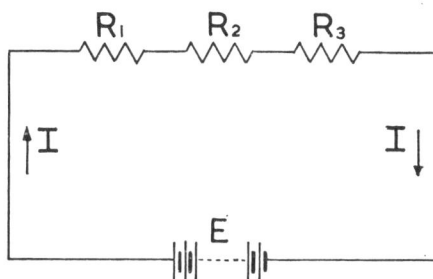

FIGURE 2.1   Resistors Connected in Series.

The electric current, I, flows from the source of e.m.f. through each of the resistors in turn and so each component in the circuit carries the same current. A break anywhere in the circuit will interrupt the flow of electric current through all the resistors.

2.12 Each resistor in a series circuit requires an electric pressure between its terminals to force the electric current through it. This electric pressure is called the *potential difference* (abbreviation p.d., symbol V) and can be measured by connecting voltmeters as shown in Fig. 2.2. The p.d.s V_1, V_2 and V_3 may be calculated by means of Ohm's law; thus:

$$V_1 = R_1 \times I$$
$$V_2 = R_2 \times I$$
$$V_3 = R_3 \times I.$$

Experiments show that in a series circuit the sum of the p.d.s equals the e.m.f. applied to the circuit. This can be expressed by the formula:

$$E = V_1 + V_2 + V_3$$

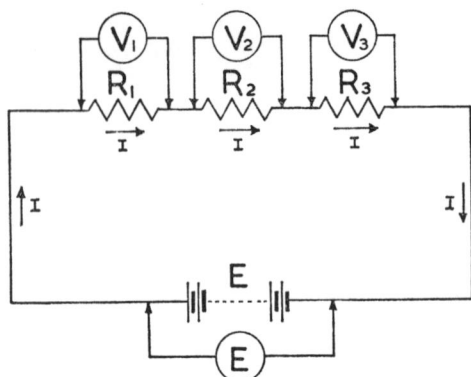

FIGURE 2.2 Measurement of p.d's and e.m.f. in a Series Circuit.

2.13 For purpose of calculation a number of resistors connected in series can be replaced by a single resistor R_T. The value of R_T can be deduced as follows.

For the circuit shown in Fig. 2.2:

$$E = V_1 + V_2 + V_3$$

By Ohm's law:

$$R_T \times I = R_1 \times I + R_2 \times I + R_3 \times I$$

Dividing throughout by I:

$$R_T = R_1 + R_2 + R_3$$

This result can be extended to any number of resistors and shows that the equivalent resistance of a number of resistors connected in series is equal to the sum of the individual resistance values. This gives the formula:

$$R_T = R_1 + R_2 + R_3 \cdots + R_n$$

Example 2.1

Two resistors of 6·2 Ω and 3·8 Ω respectively are connected in series with a 12 V battery. Determine, (a) the total resistance, (b) the current flowing, (c) the p.d. across each resistor, (d) the total power consumed and the power consumed by each resistor.

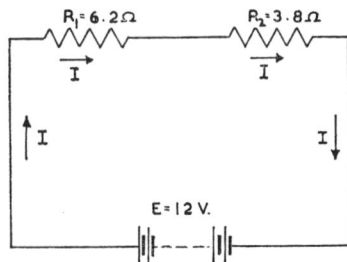

FIGURE 2.3 Example 2.1.

Solution

When solving problems concerning electric circuits it is often advantageous first to draw the circuit diagram. Figure 2.3 shows the circuit concerned in this example.

$$(a) \ R_T = R_1 + R_2$$
$$R_T = 6{\cdot}2 + 3{\cdot}8$$
$$\therefore \quad \underline{R_T = 10 \ \Omega}$$

$$(b) \ I = \frac{E}{R_T}$$
$$I = \frac{12}{10}$$
$$\underline{I = 1.2 \ A}$$

(c) p.d. across $R_1 = V_1$ and by Ohm's law:
$$V_1 = R_1 \times I$$
$$V_1 = 6{\cdot}2 \times 1{\cdot}2$$
$$\underline{V_1 = 7{\cdot}44 \ V}$$

p.d. across $R_2 = V_2$
$$V_2 = R_2 \times I$$
$$V_2 = 3{\cdot}8 \times 1{\cdot}2$$
$$\underline{V_2 = 4{\cdot}56 \ V}$$

As a useful check on the working:
$$E = V_1 + V_2$$
$$E = 7{\cdot}44 + 4{\cdot}56$$
$$\underline{E = 12 \ V} \text{ which is correct.}$$

(d) Total power consumed:
$$P_T = E \times I$$
$$P_T = 12 \times 1{\cdot}2$$
$$\underline{P_T = 14{\cdot}4 \ W}$$

Power consumed by R_1:
$$P_1 = V_1 \times I$$
$$P_1 = 7{\cdot}44 \times 1{\cdot}2$$
$$\underline{P_1 = 8{\cdot}93 \ W}$$

Power consumed by R_2:

$$P_2 = V_2 \times I$$
$$P_2 = 4\cdot56 \times 1\cdot2$$
$$P_2 = 5\cdot47 \text{ W}$$

Check:

$$P_T = P_1 + P_2$$
$$P_T = 8\cdot93 + 5\cdot47$$
$$P_T = 14\cdot4 \text{ W which is correct.}$$

N.B. The answers to example 2.1(d) have been calculated to 3 significant figures.

2.14 Cells may be connected in series in order to provide a higher e.m.f. than that available from one cell alone. The total e.m.f. of a number of cells in series is equal to the sum of the individual e.m.f.s, while the total internal resistance is the sum of the internal resistances. Thus for n cells, each of e.m.f. E volts and internal resistance r ohms connected in series, the total e.m.f. is:

$$E_T = n \times E$$

and the total internal resistance is:

$$r_T = n \times r.$$

Example 2.2

Describe the construction of a Leclanché Cell.

A battery consists of five cells, each having an e.m.f. of 1·2 volts and an internal resistance of 0·4 ohms, joined in series. If this battery supplies current to a bell of resistance 6 ohms, calculate the current flowing and the p.d. across the battery terminals. (W.J.E.C.)

Solution

For the description of a Leclanché Cell see Chapter 1, Section 1.72.

Figure 2.4 shows the circuit diagram for this example, it should be

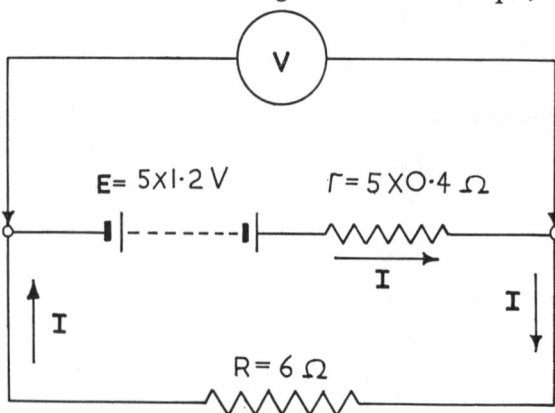

FIGURE 2.4   Example 2.2.

noted that the internal resistance is represented theoretically as an additional series resistor.

Total e.m.f. for five cells in series $= 5 \times 1 \cdot 2$

$$E = 6 \text{ V}$$

Total internal resistance for five cells in series $= 5 \times 0 \cdot 4$

$$r = 2 \, \Omega$$

$$R_T = r + R_1$$
$$R_T = 2 + 6$$
$$R_T = 8 \, \Omega$$
$$I = \frac{E}{R_T}$$
$$I = \frac{6}{8}$$
$$\underline{I = 0 \cdot 75 \text{ A}}$$

P.D. across battery terminals $= E - rI$

i.e. $V = 6 - 2 \times 0 \cdot 75$

whence $V = 4 \cdot 5 \text{ V}$

2.2 Parallel Circuits

2.21 In a parallel circuit the connections are such that the same value of electric pressure is applied to each component. Figure 2.5 shows three

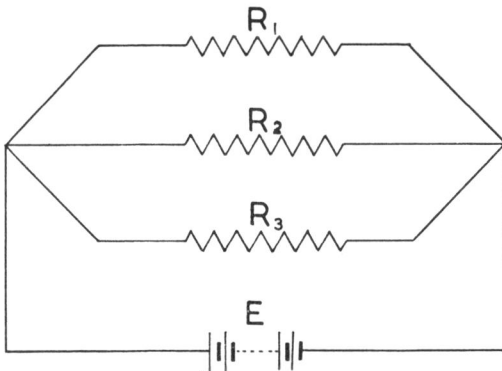

FIGURE 2.5 Resistors Connected in Parallel.

resistors connected in parallel with a source of e.m.f. (E). The full electric pressure produced by the source of e.m.f. is applied to each individual resistor, and any resistor can be disconnected without affecting the remaining resistors which will still receive the full e.m.f. of the source.

2.22 As each resistor in the parallel circuit receives the full electric pressure of the source, the electric currents can be calculated independently thus:

$$I_1 = \frac{E}{R_1}$$

$$I_2 = \frac{E}{R_2}$$

$$I_3 = \frac{E}{R_3}$$

These currents can be measured using ammeters as shown in Fig. 2.6. Experiments show that the total electric current supplied by the source

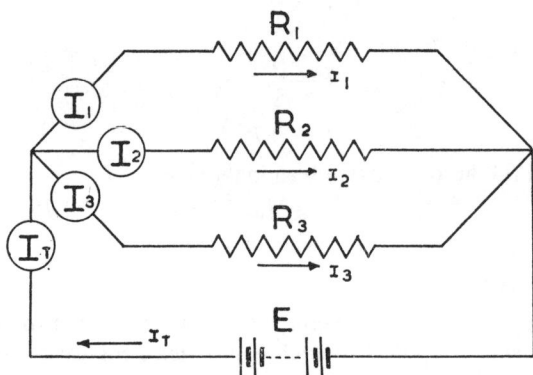

FIGURE 2.6   Measurement of Electric Currents in a Parallel Circuit.

of e.m.f. is equal to the sum of the branch currents. This can be expressed by the formula:

$$I_T = I_1 + I_2 + I_3$$

2.23 For purposes of calculation a number of resistors connected in parallel can be replaced by a single resistor R_T. The value of R_T can be deduced as follows:

For the circuit shown in Fig. 2.6

$$I_T = I_1 + I_2 + I_3$$

by Ohm's law

$$\frac{E}{R_T} = \frac{E}{R_1} + \frac{E}{R_2} + \frac{E}{R_3}$$

dividing throughout by E:

$$\frac{1}{R_T} = \frac{1}{R_1} + \frac{1}{R_2} + \frac{1}{R_3}$$

This result can be extended to any number of resistors and shows that the reciprocal of the equivalent resistance of a number of resistors connected in parallel is equal to the sum of the reciprocals of the individual resistance values. This gives the formula:

$$\frac{1}{R_T} = \frac{1}{R_1} + \frac{1}{R_2} + \frac{1}{R_3} + \cdots + \frac{1}{R_n}$$

Example 2.3

Three resistors of 12 Ω, 6 Ω and 4 Ω respectively are connected in parallel with a 12 V battery. Determine (a) the total resistance and hence find the current supplied by the battery, (b) the current consumed by each resistor and check that the sum of the branch currents equals the battery current.

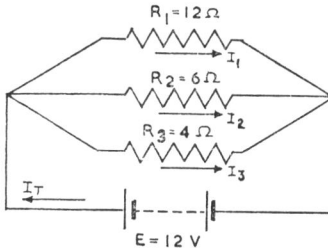

FIGURE 2.7   Example 2.3.

(a)
$$\frac{1}{R_T} = \frac{1}{R_1} + \frac{1}{R_2} + \frac{1}{R_3}$$

$$\frac{1}{R_T} = \frac{1}{12} + \frac{1}{6} + \frac{1}{4}$$

$$\frac{1}{R_T} = \frac{1 + 2 + 3}{12}$$

$$\frac{1}{R_T} = \frac{6}{12}$$

$$R_T = \frac{12}{6}$$

$$\underline{R_T = 2\ \Omega}$$

hence:

$$I_T = \frac{E}{R_T}$$

$$I_T = \frac{12}{2}$$

$$\underline{I_T = 6 \text{ A}}$$

(b)
$$I_1 = \frac{E}{R_1}$$

$$I_1 = \frac{12}{12}$$

$$\underline{I_1 = 1 \text{ A}}$$

$$I_2 = \frac{E}{R_2}$$

$$I_2 = \frac{12}{6}$$

$$\underline{I_2 = 2 \text{ A}}$$

$$I_3 = \frac{E}{R_3}$$

$$I_3 = \frac{12}{4}$$

$$\underline{I_3 = 3 \text{ A}}$$

Check:

$$I_T = I_1 + I_2 + I_3$$

$$I_T = 1 + 2 + 3$$

$$\underline{I_T = 6 \text{ A}} \text{ which is correct.}$$

2.24 Cells may be connected in parallel in order to provide a higher current than is available from one cell alone. The total e.m.f. of n cells, each of e.m.f. E volts and internal resistance r ohms is the same as the e.m.f. of one cell alone, but the internal resistance is reduced to r/n ohms and the current output capability is n times that of one cell.

Example 2.4

A battery consists of four cells, each having an e.m.f. of 2 V and an internal resistance 0·2 Ω connected in parallel. Determine the e.m.f. and internal resistance of the battery.

Solution

The e.m.f. of the battery equals the e.m.f. of one cell and so:

$$E = 2 \text{ V}$$

The internal resistance of the battery, r_T, is given by:

$$r_T = \frac{r}{n}$$

$$r_T = \frac{0\cdot2}{4}$$

$$\underline{r_T = 0\cdot05 \text{ } \Omega}$$

2.3. Combination Circuits

2.31 Many electrical circuits consist of a combination of series and parallel connections. The following examples show how typical combination circuit problems can be solved.

Example 2.5

Two resistances in parallel, *A* of 20 ohms and *B* of unknown value, are connected in series with a third resistance *C* of 12 ohms. The supply to the circuit is direct current. If the potential difference across the ends of resistance *C* is 180 volts and the power in the complete circuit is 3600 watts, calculate:

(a) The value of resistance *B*.
(b) The current in each resistance.
(c) The circuit voltage.

(C.G.L.I.)

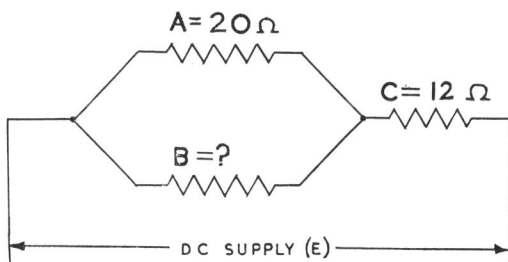

FIGURE 2.8   Example 2.5.

Solution	Notes

Solution

(i) $I_C = \dfrac{V_C}{R_C}$

$I_C = \dfrac{180}{12}$

$I_C = 15$ A

(ii) $P = R \times I^2$

$R_T = \dfrac{P}{I^2}$

$R_T = \dfrac{3600}{15 \times 15}$

$R_T = 16\ \Omega$

(iii) $R_{AB} = R_T - R_C$

$R_{AB} = 16 - 12$

$R_{AB} = 4\ \Omega$

(iv) $V_{AB} = R_{AB} \times I_T$

$V_{AB} = 4 \times 15$

$V_{AB} = 60$ V

(v) $I_A = \dfrac{V_A}{R_A}$

$I_A = \dfrac{60}{20}$

$I_A = 3$ A

(vi) $I_B = I_T - I_A$

$I_B = 15 - 3$

$I_B = 12$ A

(vii) $R_B = \dfrac{V_B}{I_B}$

$R_B = \dfrac{60}{12}$

$R_B = 5\ \Omega$

(viii) $E = V_{AB} + V_C$

$E = 180 + 60$

$E = 240$ V

Notes

(i) As both the p.d. and resistance of C are known, the current can be calculated using Ohm's law. There are no resistors connected in parallel with C therefore this current is also the total current in the circuit.

(ii) As both the total power and current in the circuit are known, the total resistance can be calculated, using the power formula $P = R \times I^2$.

(iii) The *combined* resistance of A and B is in series with C and so $R_T = R_{AB} + R_C$.

(iv) Now that the combined resistance of A and B and the total current are known, the p.d. across the combination of A and B can be calculated using Ohm's law.

(v) Again Ohm's law is used; noting that as A and B are connected in parallel, the p.d.s V_A and V_B across A and B are both equal to V_{AB}.

(vi) The total current in a parallel circuit is the sum of the branch currents.

(vii) Now that V_B and I_B are known Ohm's law is used to calculate R_B.

(viii) The total e.m.f. equals the sum of the p.d.s connected in series.

Answers:

(a) $B = 5\ \Omega$

(b) $I_A = 3$ A $I_B = 12$ A $I_C = 16$ A

(c) $E = 240$ V

There are often several alternative ways of solving a given problem of this type. For example, in the above problem the value of E could have been calculated in step (ii) using the formula $E = P/I$. Then V_{AB} could be found from the formula $V_{AB} = E - V_C$, the solution then proceeding as given from step (v) onwards.

Example 2.6

Three resistances of value 1·5 ohms, 4 ohms and 12 ohms respectively are connected in parallel. A fourth resistance of 6 ohms is connected in series with the parallel group. A d.c. supply of 140 volts is applied to the circuit.

(a) Calculate the current taken from the supply.
(b) Find the value of a further resistance to be connected in parallel with the 6 ohms resistance, so that the potential difference across it shall be 84 volts.
(c) What current will now flow in the circuit?   (C.G.L.I.)

Solution

Let A denote the parallel group of resistors and B the single resistor in series.

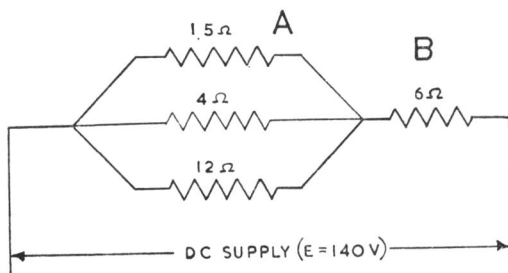

FIGURE 2.9   Example 2.6(a).

Calculations

(i)
$$\frac{1}{R_A} = \frac{1}{R_1} + \frac{1}{R_2} + \frac{1}{R_3}$$
$$\frac{1}{R_A} = \frac{1}{1\cdot5} + \frac{1}{4} + \frac{1}{12}$$
$$\frac{1}{R_A} = \frac{8+3+1}{12}$$
$$\frac{1}{R_A} = \frac{12}{12}$$
$$R_A = \frac{12}{12}$$
$$R_A = 1\ \Omega$$

Notes

(i) Use the formula for resistors in parallel to find the total resistance of group A

29

(ii) $R_T = R_A + R_B$
$R_T = 1 + 6$
$R_T = 7\ \Omega$

(ii) Use the formula for resistors in series to find the total resistance of the circuit.

(iii) $I = \dfrac{E}{R_T}$

$I = \dfrac{140}{7}$

$I = 20\ A$

(iii) The total current can now be found by applying Ohm's law.

This answers part (a) of the question. A new circuit diagram, Fig. 2.10 is required to complete the remainder of the question. In this diagram a

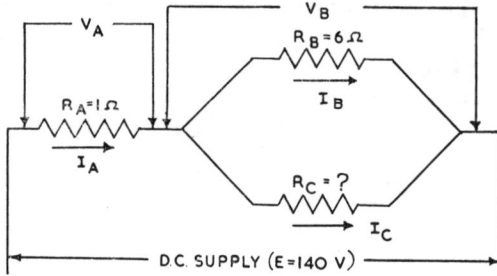

FIGURE 2.10   Example 2.6(b)

single resistor R_A can be shown as the equivalent of the three original resistors of section A of the circuit. The value of R_A has already been found to be 1 ohm in section (i).

Calculations

(iv) $V_A = E - V_B$
$V_A = 140 - 84$
$V_A = 56\ V$

(v) $I_A = \dfrac{V_A}{R_A}$

$I_A = \dfrac{56}{1}$

$I_A = 56\ A$

(vi) Current through 6 Ω resistor $= I_B$.

$I_B = \dfrac{V_B}{R_B}$

$I_B = \dfrac{84}{6}$

$I_B = 14\ A$

Notes

(iv) Use the rule "p.d.s in series add". i.e. $E = V_A + V_B$

(v) Apply Ohm's law to section A. Note that I_A is also the total current flowing in the circuit.

(vi) Apply Ohm's law to resistor R_B.

(vii) $I_C = I_A - I_B$
$I_C = 56 - 14$
$I_C = 42$ A

(vii) Use the rule "currents in parallel add". i.e. $I_A = I_B + I_C$

(viii) $R_C = \dfrac{V_C}{I_C}$

$R_C = \dfrac{84}{42}$

$R_C = 2\ \Omega$

(viii) Apply Ohm's law to calculate R_C.

Answers: (a) 20 A
(b) 2 Ω
(c) 56 A

2.4. Wheatstone's Bridge

2.41 Wheatstone's bridge circuit which is illustrated by Fig. 2.11(a) provides a useful arrangement for the accurate measurement of resistance.

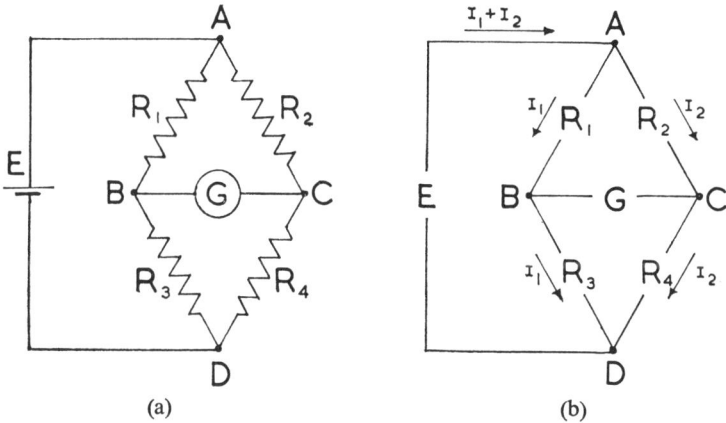

FIGURE 2.11 Wheatstone Bridge.
(a) Basic circuit.
(b) Currents in balanced bridge.

One or more of the resistors forming the bridge circuit can be varied, and it is possible to adjust the resistors so that there is no potential difference between terminals *B* and *C*. The bridge is then said to be *balanced* and, as there is no p.d. between the galvanometer terminals, the balanced condition is indicated by a zero reading on the galvanometer. When the bridge is balanced none of the current I_1 flowing through resistor R_1 is diverted through the galvanometer. Since, under balanced conditions, this instrument receives no p.d. and hence carries no current, it follows that *all* of the current I_1 flows through resistor R_3. Similarly all of the current

31

I_2 flowing through resistor R_2 must also flow through resistor R_4. These currents are marked on the diagram in Fig. 2.11(b). Terminal D can be assumed to be at a potential of zero volts and, in this case:

The potential at B can be found by applying Ohm's law to resistor R_3

$$V_B = R_3 \times I_1$$

Similarly the potential at C is given by:

$$V_C = R_4 \times I_2$$

But there is no potential difference between B and C

$$\therefore \quad V_B = V_C$$

$$\text{i.e.} \quad R_3 \times I_1 = R_4 \times I_2 \qquad \text{....(i)}$$

The potential at A is maintained at E volts by the battery so the potential at B can also be found by applying Ohm's law to resistor R_1:

$$V_B = E - R_1 \times I_1$$

$$\text{Similarly:} \quad V_C = E - R_2 \times I_2$$

$$\text{Again:} \quad V_B = V_C$$

$$\therefore \quad E - R_1 \times I_1 = E - R_2 \times I_2$$

$$R_1 \times I_1 = R_2 \times I_2 \qquad \text{....(ii)}$$

By combining equations (i) and (ii)

$$\frac{R_1 \times I_1}{R_3 \times I_1} = \frac{R_2 \times I_2}{R_4 \times I_2}$$

$$\frac{R_1}{R_3} = \frac{R_2}{R_4}$$

2.42 The equation derived above shows that when the bridge is balanced, the resistors must be in a definite ratio to each other so that, if the values of three resistors are known, the value of the fourth may be easily calculated. The value of the e.m.f. (E) which energises the circuit does not affect the conditions of balance; the e.m.f. used need only be large enough to produce a satisfactory deflection on the galvanometer when the bridge is unbalanced. By using a sensitive galvanometer the balance point can be found very precisely and, as accurately calibrated resistors are readily available, the wheatstone bridge circuit can provide an extremely accurate method of resistance measurement.

2.43 A simply constructed form of wheatstone bridge is the *slide-wire bridge* which is illustrated in Fig. 2.12. The slide wire bridge is usually constructed on a wooden baseboard, and consists of three copper bus-bars (labelled *a*, *b* and *c* in Fig. 2.12). A standard resistor R_s of known value is connected between *a* and *b* while the unknown resistor R_x is connected between *b* and *c*. A uniform length of resistance wire is stretched tautly between *a* and *c*, and in many bridges the length of this wire is 1 metre. The galvanometer, which is used to indicate balanced conditions, is connected between *b* and a sliding contact *d* which can be moved to any desired position along the resistance wire in order to find the balance point. A scale is fitted so that the lengths *s* and *x* of resistance wire on either side of the slider can be measured. The bridge is energised by

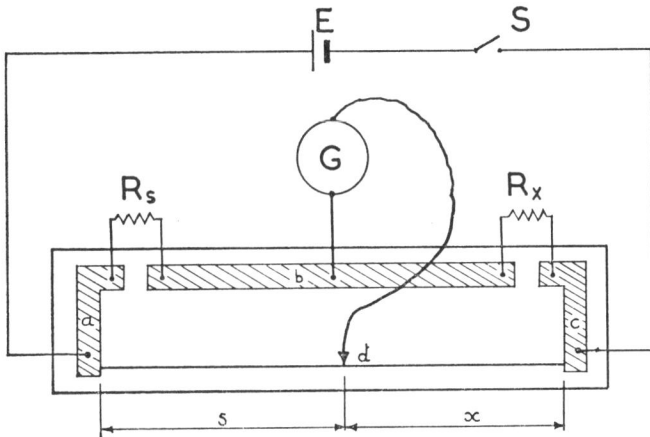

FIGURE 2.12   Slide Wire Bridge.

connecting a source of e.m.f. (*E*) between *a* and *c*. By comparing the layout of the slide bridge with the basic wheatstone bridge circuit of Fig. 2.11 it can be seen that:

R_1 of the basic circuit corresponds with R_s of the slide wire bridge. Similarly R_2 corresponds with R_x, R_3 with the length of resistance wire *s* and R_4 with the length of resistance wire *x*. For the basic circuit:

$$\frac{R_1}{R_3} = \frac{R_2}{R_4}$$

$$\therefore \quad R_2 = R_1 \times \frac{R_4}{R_3}$$

In the case of the slide wire bridge the ratio of the resistances R_4/R_3 equals the ratio of the lengths x/s since, for a uniform wire, resistance is proportional to length. Thus the equation for the slide-wire bridge becomes

$$R_x = R_s \times \frac{x}{s}$$

2.44   There are many practical compact forms of wheatstone bridge available. In some of these bridges the slide wire is in the form of a circular

33

rheostat fitted with a dial calibrated to read resistance values directly, and a range switch is provided to select the required value of standard resistor. In other types of bridge the range switch selects suitable pairs of resistors (R_4 and R_3) to form a *ratio arm* and the standard resistor (R_1) is in the form of a resistance decade; i.e. there are a number of ten-position switches, the first giving values in one ohm steps from 1 to 10 ohms, the next giving values in ten ohm steps from 10 to 100 ohms, and so on. In all of these bridges the scales are arranged to read ohms directly and the range switch is often designed to give ranges in the sequence: times 1; times 10; times 100 etc., so that calculations are reduced to a minimum.

Example 2.7

The wheatstone bridge circuit shown in Fig. 2.13 was used to measure the resistance of a field coil. R_1 and R_2 were fixed resistors of value 10 ohms

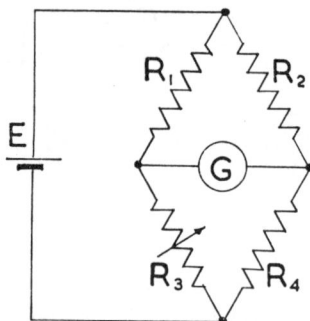

FIGURE 2.13   Example 2.7.

and 100 ohms respectively, R_3 was a decade resistance and R_4 the field coil. If balance was obtained when R_3 was set to 24·9 ohms, what was the resistance of the field coil?

Solution

$$\frac{R_1}{R_3} = \frac{R_2}{R_4}$$

$$R_4 = \frac{R_3 \times R_2}{R_1}$$

$$R_4 = \frac{24\cdot9 \times 100}{10}$$

$$\underline{R_4 = 249\ \Omega}$$

Example 2.8

The slide wire bridge circuit shown in Fig. 2.14 was balanced when $x = 340$ mm and $s = 660$ mm. R_s had a value of 25 ohms, calculate the value of R_x.

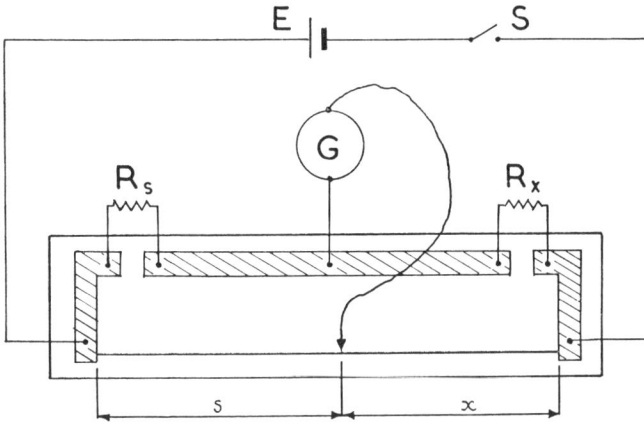

FIGURE 2.14 Example 2.8.

Solution

$$R_x = R_s \times \frac{x}{s}$$

$$= 25 \times \frac{340}{660}$$

$$\underline{R_x = 12 \cdot 9 \ \Omega}$$

SUMMARY OF FORMULAE

Resistors in series: $\qquad R_T = R_1 + R_2 + R_3 + \cdots R_n$

Resistors in parallel: $\qquad \dfrac{1}{R_T} = \dfrac{1}{R_1} + \dfrac{1}{R_2} + \dfrac{1}{R_3} + \cdots \dfrac{1}{R_n}$

Cells in series: $\qquad E_T = nE$

$$r_T = nr$$

Cells in parallel: $\qquad r_T = \dfrac{r}{n}$

Wheatstone bridge: $\qquad \dfrac{R_1}{R_3} = \dfrac{R_2}{R_4}$

EXERCISE 2

1. Given three resistors R_1, R_2, R_3, of values 20 ohms, 40 ohms and 60 ohms respectively; what values of resistance can be obtained by
   (a) connecting any two in series?
   (b) connecting all three in series?
   (c) connecting any two in parallel?
   (d) connecting all three in parallel?

2. An electric hotplate has two heating elements each of resistance 60 ohms. The hotplate is controlled by a "three heat" switch which:
   (a) in the LOW position connects both elements in series with the supply,
   (b) in the MEDIUM position connects one element only to the supply,
   (c) in the HIGH position connects both elements in parallel with the supply.
   If the supply pressure is 240 volts calculate the power consumed by the hotplate for each position of the switch.

3. Two resistances of 10 ohms and 40 ohms respectively are connected in parallel. A third resistance of 5 ohms is connected in series with the combination, and a d.c. supply of 240 volts is applied to the ends of the complete circuit.
   (a) Calculate the current in each resistance.
   (b) What power would be spent in a fourth resistance of 20 ohms connected in parallel with the 5 ohms resistance.   C.G.L.I.

4. Three resistances of 18 ohms, 20 ohms and 30 ohms respectively are connected in parallel with each other. Two more resistors, 3·6 ohms each are also connected in parallel. The two groups of resistances are then connected in series to a 45 volt battery. Draw a circuit diagram and calculate (a) the total current, (b) the current in each resistor, the power absorbed in each of the 3·6 ohms resistors, and the potential across the 20 ohms resistor.   E.M.E.U.

5. State Ohm's law in your own words, and express it in symbols. A d.c. supply at 240 volts is applied to a circuit comprising two resistances A and B in parallel, of 5 ohms and 7·5 ohms respectively, in series with a third resistance C of 30 ohms. Calculate the value of a fourth resistance D to be connected in parallel with C, so that the total power in the circuit shall be 7·2 kW.   C.G.L.I.

6. Explain how to measure the resistance of a coil by two of the following:
   (a) slide wire bridge,
   (b) ammeter and voltmeter, and
   (c) any other method.   C.G.L.I.

7. (a) With the aid of a diagram of connections show how the value of an unknown resistance may be determined by means of a slide-wire bridge.
   (b) If the slide wire is 1 metre long and balance is obtained when the moving contact is 300 mm from the end of the wire to which is connected a standard resistance of 5 ohms, determine the value of the unknown resistance.   U.E.I.

8. Three resistances of value 3 ohms, 8 ohms and 24 ohms respectively are connected together in parallel. A fourth resistance of 5 ohms is connected in series with the parallel group. A d.c. supply at 70 V is applied to the circuit.
   (a) Calculate the current taken from the supply.
   (b) Find the value of a further resistance to be connected in parallel with the 5 ohms resistance, so that the power in the whole circuit shall be 1 508 watts.

   C.G.L.I.

9. Four resistances, AB, 3 ohms; BC, 4 ohms; CD, 6 ohms, and DA, 6 ohms, are connected in series to form a closed square $ABCD$.
   A d.c. supply at 70 V is connected across A and C, so that current flows into the arrangement at A. A high resistance voltmeter is connected between B and D.
   (a) Find the reading of the voltmeter, and give the direction of the voltmeter current.
   (b) Find the value of a further resistance X, to be connected in parallel with DA, so that the voltmeter reads zero.
   (c) Find the total current flowing in the circuit in case (b).   C.G.L.I.

10. A direct-current circuit comprises two resistors, A of value 25 ohms, and B of unknown value, connected in parallel, together with a third resistor C of value 5 ohms connected in series with the parallel group. The potential difference across C

is found to be 90 V. If the total power in the circuit is 4320 W, calculate:
(a) the value of resistor B,
(b) the voltage applied to the ends of the whole circuit,
(c) the current in each resistor.   C.G.L.I.

11. A battery of 9 primary cells is connected,
    (a) all cells in series
    (b) all cells in parallel
    (c) three sets in parallel, each set consisting of 3 cells in series.
    Each cell has e.m.f. of 1·4 volts, and internal resistance of 0·45 ohm. The battery terminals are connected to a circuit of resistance 7·2 ohms. Calculate in each case,
    (a) the current in the 7·2 ohm resistance
    (b) the voltage drop across the resistance.   C.G.L.I.

12. Describe, with sketches, a lead acid secondary cell, and state briefly the chemical changes in the cell during charge and discharge. (The chemical formulae are *not* required.) Explain the importance of a low internal resistance.
    A lead acid cell discharging at the rate of 6 amperes has a terminal potential difference of 1·95 volts. On open circuit the potential difference is 2·1 volts. Calculate the value of the internal resistance of the cell.   C.G.L.I.

3 Electric Circuits II

3.1. Conductor Resistance

3.11 Every conductor possesses resistance, the value of which depends on four factors

(a) the type of conducting material (M)
(b) the cross-sectional area (A) of the conductor
(c) the length (L) of the conductor
(d) the temperature (T)

The word *MALT* provides a useful aid to the memory when dealing with these factors.

3.12 A conductor of length L can be regarded as being made up of a number of short conductors joined in series as shown in Fig. 3.1. If the lengths of the short conductors are l_1, l_2, l_3 etc. then the total length L made up of n lengths is given by:

$$L = l_1 + l_2 + l_3 + \cdots l_n.$$

Also, if the resistances of the short conductors are r_1, r_2, r_3 etc., then the total resistance R is given by

$$R = r_1 + r_2 + r_3 + \cdots r_n.$$

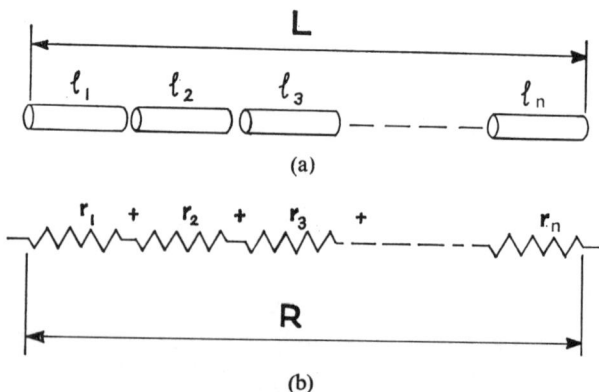

(a)

(b)

FIGURE 3.1 Conductor Resistance Proportional to Length.

If all the short conductors have the same length l_1, they will have the same resistance r_1, and in this case:

$$L = nl_1$$

and
$$R = nr_1$$

Whence
$$\frac{R}{L} = \frac{nr_1}{nl_1}$$

i.e.
$$R = \left(\frac{r_1}{l_1}\right)L$$

But (r_1/l_1) is a constant and so it follows that the resistance of a conductor is directly proportional to its length.

3.13 A conductor of cross-sectional area A can be regarded as being made up of a number of smaller conductors joined in parallel as shown in Fig. 3.2. If the areas of the small conductors are a_1, a_2, a_3 etc., then the total area A made up of n small conductors is given by:

$$A = a_1 + a_2 + a_3 + \cdots a_n$$

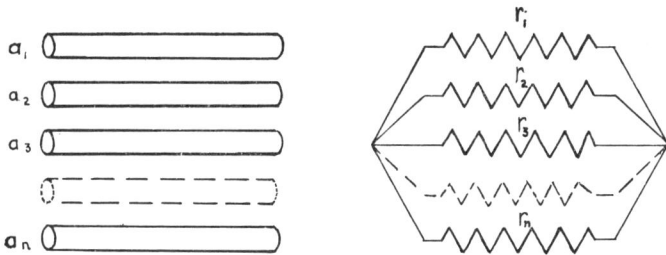

FIGURE 3.2 Conductor Resistance Inversely Proportional to Cross-Sectional-Area.

Also, if the resistances of the small conductors are r_1, r_2, r_3, etc., then the total resistance R is given by:

$$\frac{1}{R} = \frac{1}{r_1} + \frac{1}{r_2} + \frac{1}{r_3} + \cdots \frac{1}{r_n}$$

If all the small conductors have the same area a_1, they will have the same resistance r_1, and in this case

$$A = na_1$$

and
$$\frac{1}{R} = \frac{n}{r_1} \quad \text{whence } R = \frac{r_1}{n}$$

But
$$RA = \frac{r_1}{n} \times na_1$$

Whence
$$R = (r_1 a_1) \times \frac{1}{A}$$

But $(r_1 a_1)$ is a constant and so it follows that the resistance of a conductor is inversely proportional to its cross-sectional area.

ELECTRICAL PRINCIPLES

Example 3.1

If the resistance of 100 m of a certain conductor is 1·24 Ω, calculate the resistance of 500 m of conductor of the same material but of twice the cross-sectional area.

Solution

Let R_1 = first value of resistance and R_2 second value of resistance,

l_1 = first length and l_2 = second length,

A_1 = first cross-sectional area and A_2 = second cross-sectional area.

Then as conductor resistance is directly proportional to length and inversely proportional to cross-sectional area:

$$R_2 = R_1 \times \frac{l_2}{l_1} \times \frac{A_1}{A_2}$$

$$R_2 = 1\cdot24 \times \frac{500}{100} \times \frac{1}{2}$$

$$\underline{R_2 = 3\cdot1 \ \Omega}$$

3.2. Insulation Resistance

3.21 The *insulation resistance* of a cable may be defined as the opposition offered by the insulation to the leakage of current from the cable. As is shown in Fig. 3.3 the length *of insulation* through which the leakage

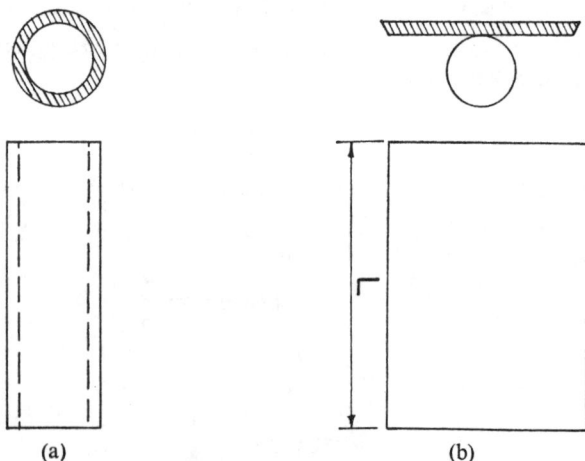

(a)   (b)

FIGURE 3.3   Insulation Resistance.
(a) Insulation around Conductor.
(b) Insulation opened out to show effective cross-sectional area.

40

current flows is equal to the thickness of the insulation and is constant, but the cross-sectional area *of insulation* is proportional to the length of the cable. It follows that, since resistance is inversely proportional to cross-sectional area, the insulation resistance of a cable is inversely proportional to its length.

Example 3.2

Calculate the insulation resistance of 200 m of insulated cable if the insulation resistance of 1000 m is 1500 megohms.

Solution

Insulation resistance is inversely proportional to length of cable so for 200 m of the cable

$$\text{Insulation resistance} = 1500 \times \frac{1000}{200}$$

$$= 7500 \text{ M}\Omega$$

3.3. Resistivity

3.31 The resistivity (symbol ρ) of a material is defined as the resistance of a sample of unit length and unit cross-sectional area. A material which is a good conductor has a low value of resistivity while a poorer conductor has a higher resistivity. The table below gives the resistivities of some commonly used conductors.

Material	Resistivity at 0°C
Silver	$16\cdot5 \times 10^{-9}$ ohm metre
Copper	$17\cdot3 \times 10^{-9}$ ohm metre
Aluminium	$28\cdot4 \times 10^{-9}$ ohm metre
Carbon	$70\,000 \times 10^{-9}$ ohm metre
Nichrome*	$1\,110 \times 10^{-9}$ ohm metre
Eureka*	490×10^{-9} ohm metre

* The exact value depends on the composition of the alloy.

The table lists the resistivity of copper as $17\cdot3 \times 10^{-9}$ ohm metre. This means that the resistance of a sample of copper one metre long and one square metre in cross-sectional area is $0\cdot000\,000\,017\,3$ ohm, that is $\frac{17\cdot3}{1\,000\,000\,000}$ ohm.

3.32 The resistance of any conductor is directly proportional to its length and inversely proportional to its cross-sectional area. Now for

41

$l_1 = 1$ metre and $A_1 = 1$ square metre, the resistance $r_1 = \rho$. Thus for $l_2 = l$ metres and $A_2 = A$ square metres the resistance R is given by

$$R = \rho \times \frac{l}{1} \times \frac{1}{A}$$

$$R = \frac{\rho l}{A}$$

Example 3.3

Calculate the resistance of a copper wire 100 m long and 1·5 mm diameter. The resistivity of copper is $17\cdot3 \times 10^{-9}\ \Omega$ m.

Solution

$$\text{Cross-sectional area of conductor} = \frac{\pi d^2}{4}$$

$$A = \frac{3\cdot142 \times 1\cdot5 \times 1\cdot5}{4}$$

$$A = 1\cdot767\ \text{mm}^2$$

In order to calculate the resistance the area is required to be expressed in m^2 and so:

$$A = 1\cdot767 \times 10^{-6}\ \text{m}^2$$

or

$$A = \frac{1\cdot767}{1\ 000\ 000}\ \text{m}^2$$

$$R = \frac{\rho l}{A} = \rho \times l \times \frac{1}{A}$$

$$R = \frac{17\cdot3}{1\ 000\ 000\ 000} \times 100 \times \frac{1\ 000\ 000}{1\cdot767}$$

$$\underline{R = 0\cdot979\ \Omega}$$

3.4. Voltage Drop

3.41 When a resistor is connected in series with a load, the electric pressure across the load is less than that available at the supply terminals. The difference is due to the p.d. which is required to force the electric current through the resistor. This p.d. is called the voltage drop across the resistor and it can easily be calculated using Ohm's law, $(V = R \times I)$.

Example 3.4

A slide projector uses a bulb rated at 200 W, 100 V. A voltage dropping resistor is to be connected in series with the bulb so that it can be operated correctly from a 240 V supply. What value resistor is required?

Solution

$$\text{Current consumed by bulb} = \frac{W}{V}$$

$$I = \frac{200}{100} = 2 \text{ A}$$

$$\text{Voltage drop required} = 240 - 100$$

$$V = 140 \text{ V}$$

$$\therefore \quad \text{Resistance value required} = \frac{V}{I} = \frac{140}{2}$$

$$\underline{R = 70 \ \Omega}$$

3.42 As every conductor possesses resistance, voltage drops will always occur in the cores of the cables joining a source of supply to its load, so that the p.d. across the load is always less than the e.m.f. of the supply. Figure 3.4 shows that, in a simple supply circuit, the current flows

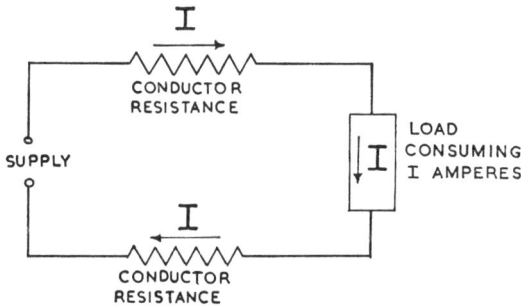

FIGURE 3.4 Simple Supply Circuit.

to the load via the resistance of one conductor and returns via the resistance of a second conductor. Thus the total series resistance causing a voltage drop is the sum of the resistance of the two conductors. By applying Ohm's law the p.d. across the load is given by:

V (p.d. across load) = *E* (supply pressure) − *R* (total conductor resistance) × *I* (current in conductor).

3.43 The passage of the load current through the resistance of the conductors causes heat to be produced. The power lost due to this effect can be found by applying the formula:

Power lost in cable = Total cable resistance × square of load current.

i.e. $$P = R \times I^2$$

Example 3.5

A load of 150 amperes is supplied from a 240 volt sub-station by means of a twin-cable, each cable core having a resistance of 0·02 ohms. Calculate the p.d. across the load and the power loss in the cable.

Solution

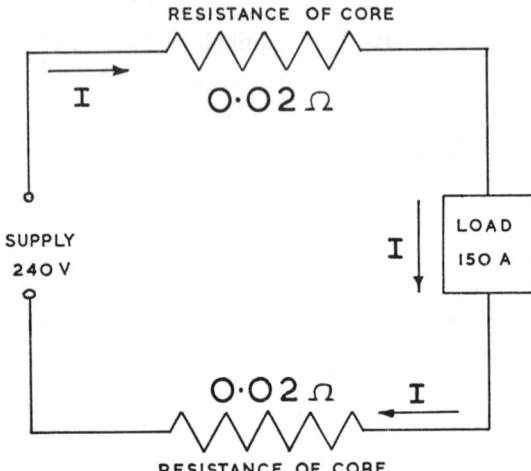

RESISTANCE OF CORE

I

$0·02\,\Omega$

SUPPLY
240 V

I

LOAD
150 A

$0·02\,\Omega$ I

RESISTANCE OF CORE

FIGURE 3.5 Example 3.5.

Total resistance of 2 cores $= 2 \times 0·02 = 0·04\ \Omega$

p.d. across load $= 240 - 0·04 \times 150$

$$V = 234\ \text{V}$$

Power loss $= RI^2$

$$P = 0·04 \times 150^2$$

$$P = 900\ \text{W}$$

3.5. Voltage Drop Problems

3.51 A two wire main can be used to supply several loads, as shown in Fig. 3.6.

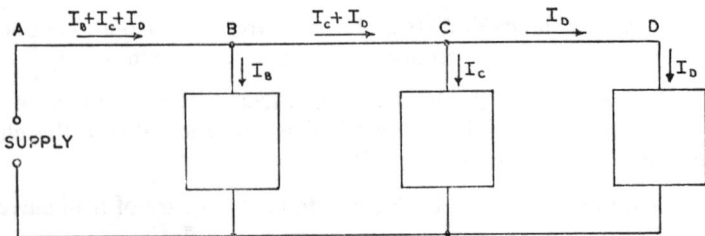

A $I_B + I_C + I_D$ B $I_C + I_D$ C I_D D

I_B I_C I_D

SUPPLY

FIGURE 3.6 Two-Wire Main Supplying Several Loads.

It can be seen from the diagram that the various sections of the cable carry different values of current. The section CD, which is the section furthest from the supply, carries only the current of the last load, I_D. Section BC must supply both loads C and D and so carries the sum of these two load currents, $I_C + I_D$. Similarly, section AB supplies loads B, C and D and so carries a current $I_B + I_C + I_D$. The voltage drop in each section of the cable must be determined separately and the total voltage up to any load found by summing the voltage drops in the sections of cable supplying the load, as illustrated in example 3.6.

Example 3.6

A two core cable is fed with a constant pressure of 250 volts at A. Loads of 50, 20 and 30 amperes are connected at points B, C and D respectively. The lengths of cable are: A to B 500 m; B to C 700 m; C to D 300 m; and each core of the cable has a resistance of 0·1 ohm per 1000 m. Draw a diagram showing the current in each section of the cable and determine the p.d. across each load.

Solution

The current distribution is as shown in Fig. 3.7, the currents in the cable sections being obtained by adding the load currents one at a time as explained in paragraph 3.51.

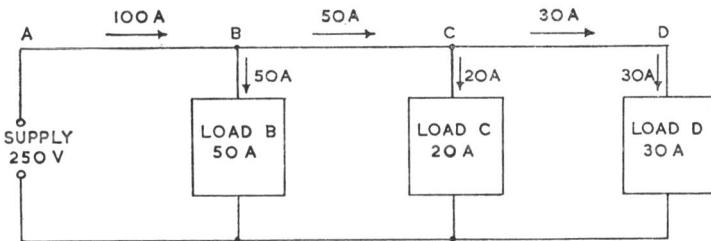

FIGURE 3.7   Example 3.6.

$$\text{Resistance of cable section } AB = 2 \times 500 \times \frac{0\cdot 1}{1000}$$

$$R_{AB} = 0\cdot 1 \ \Omega \text{ (for two cores)}$$

$$\text{Resistance of cable section } BC = 2 \times 700 \times \frac{0\cdot 1}{1000}$$

$$R_{BC} = 0\cdot 14 \ \Omega$$

$$\text{Resistance of cable section } CD = 2 \times 300 \times \frac{0\cdot 1}{1000}$$

$$R_{CD} = 0\cdot 06 \ \Omega$$

$$\text{Voltage drop in cable section } AB = R_{AB} \times I_{AB}$$
$$= 0\cdot 1 \times 100$$
$$= 10 \text{ V}$$

45

Voltage drop in cable section $BC = R_{BC} \times I_{BC}$
$$= 0{\cdot}14 \times 50$$
$$= 7\,\text{V}$$
Voltage drop in cable section $CD = R_{CD} \times I_{CD}$
$$= 0{\cdot}06 \times 30$$
$$= 1{\cdot}8\,\text{V}$$
P.D. across load $B = E - R_{AB}I_{AB}$
$$V_B = 250 - 10$$
$$\underline{V_B = 240\,\text{V}}$$
P.D. across load $C = V_B - R_{BC}I_{BC}$
$$V_C = 240 - 7$$
$$\underline{V_C = 233\,\text{V}}$$
P.D. across load $D = V_C - R_{CD}I_{CD}$
$$V_D = 233 - 1{\cdot}8$$
$$\underline{V_D = 231{\cdot}2\,\text{V}}$$

Example 3.7

A d.c. distribution cable 240 m long has two copper cores each of 100 mm² cross-sectional area and is fed at one end at the constant voltage of 250 V. A consumer A, whose full load current is 40 A is connected to the other end of the distribution cable by a service cable 20 m long having two cores each of 50 mm² cross-sectional area. Another consumer, B, whose full load current is 60 A is connected to the distribution cable at a point 100 m from connection A by a service cable 30 m long with two cores each of 50 mm² cross-sectional area. Find the potential difference at each consumer's terminals when both are taking full load. The resistivity of copper is $17{\cdot}3 \times 10^{-9}\,\Omega$ m.

Solution

First draw a diagram showing the connections and determine the current distribution. (See Figure 3.8.)

The resistance of each cable section may be calculated by using the expression:
$$R = \frac{\rho l}{A}$$

Resistance of cable section $CD = \dfrac{17{\cdot}3 \times 140 \times 2 \times 1\,000\,000}{1\,000\,000\,000 \times 100}$ (for two cores)
$$R_{CD} = 0{\cdot}0484\,\Omega$$

Resistance of cable section $DE = \dfrac{17{\cdot}3 \times 100 \times 2 \times 1\,000\,000}{1\,000\,000\,000 \times 100}$
$$R_{DE} = 0{\cdot}0346\,\Omega$$

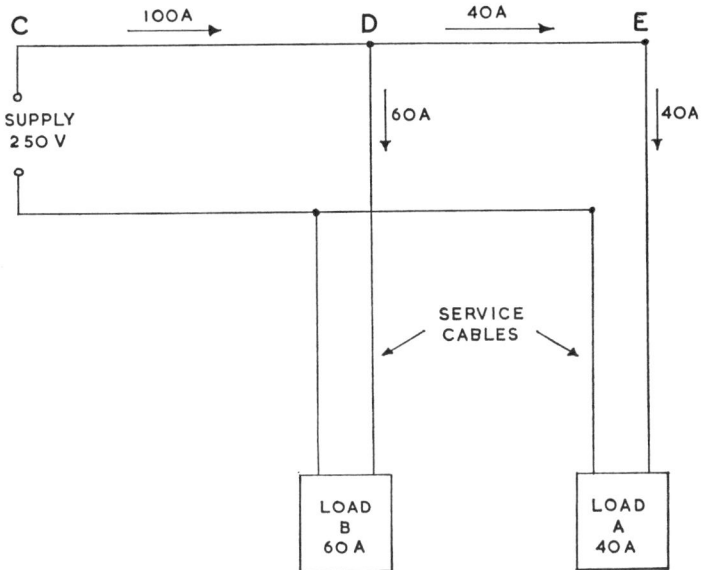

FIGURE 3.8   Example 3.7.

Resistance of cable section $DB = \dfrac{17 \cdot 3 \times 30 \times 2 \times 1\,000\,000}{1\,000\,000\,000 \times 50}$

$$R_{DB} = 0 \cdot 0208 \; \Omega$$

Resistance of cable section $EA = \dfrac{17 \cdot 3 \times 20 \times 2 \times 1\,000\,000}{1\,000\,000\,000 \times 50}$

$$R_{EA} = 0 \cdot 0138 \; \Omega$$

Voltage drop in section $CD = 0 \cdot 0484 \times 100$

$$= 4 \cdot 84 \text{ V}$$

Voltage drop in section $DE = 0 \cdot 0346 \times 40$

$$= 1 \cdot 38 \text{ V}$$

Voltage drop in section $DB = 0 \cdot 0208 \times 60$

$$= 1 \cdot 25 \text{ V}$$

Voltage drop in section $EA = 0 \cdot 0138 \times 40$

$$= 0 \cdot 55 \text{ V}$$

P.D. at junction $D = 250 - 4 \cdot 84$

$$V_D = 245 \cdot 16 \text{ V}$$

P.D. at load $B = 245 \cdot 16 - 1 \cdot 25$

$V_B = 243 \cdot 92$, (say 243·9 V to the
nearest $\frac{1}{10}$ volt)

P.D. at load $A = 245 \cdot 16 - 1 \cdot 38 - 0 \cdot 55$

$V_A = 243 \cdot 13$ (say 243·1 to the
nearest $\frac{1}{10}$ volt)

3.52 The maximum current which can be allowed to flow in any cable is limited by the heat produced in the cable conductors by the passage of the current. The amount of current which may be allowed depends on the material and cross-sectional area of the conductors, the thickness and type of insulation and on the conditions in which the cables are installed. The current loading of cables is sometimes compared by calculating the *current density*, where:

$$\text{Current Density} = \frac{\text{Current}}{\text{Cross-Sectional-Area of Conductor}}$$

Should a consumer's load requirements increase, the loading may eventually exceed the current carrying capacity of the existing main cable. In such a situation it may be preferable to install an additional main cable connected in parallel with the existing main, rather than to replace the main supply cable with one of greater current carrying capacity.

Example 3.8

A load of 230 A is supplied from a 460 V switchboard by means of a two core copper cable and a two core aluminium cable connected in parallel. The cables are each 200 m long and each core of the cables has a cross-sectional area of 50 mm².

Calculate:

(a) The voltage drop in the combined cables at full load,
(b) The separate currents in each of the two cables.

(The resistivities of copper and aluminium may be taken as $17 \cdot 3 \times 10^{-9} \, \Omega \, \text{m}$ and $28 \cdot 4 \times 10^{-9} \, \Omega \, \text{m}$ respectively.)

Solution

(a)
$$R = \frac{\rho l}{A}$$

$$\text{Resistance of copper cable} = \frac{17 \cdot 3 \times 200 \times 2 \times 1\,000\,000}{1\,000\,000\,000 \times 50} \quad \substack{\text{(for two} \\ \text{cores)}}$$

$$R_1 = 0 \cdot 138 \, \Omega$$

$$\text{Resistance of aluminium cable} = \frac{28 \cdot 4 \times 200 \times 2 \times 1\,000\,000}{1\,000\,000\,000 \times 50}$$

$$R_2 = 0 \cdot 227 \, \Omega$$

Combined resistance of both cables in parallel:

$$\frac{1}{R} = \frac{1}{R_1} + \frac{1}{R_2}$$

$$\frac{1}{R} = \frac{1}{0\cdot138} + \frac{1}{0\cdot227}$$

$$\frac{1}{R} = 7\cdot246 + 4\cdot405$$

$$\frac{1}{R} = 11\cdot651$$

$$\therefore \quad R = \frac{1}{11\cdot65}$$

$$= 0\cdot0858 \ \Omega$$

$$\text{Voltage drop} = R \times I$$

$$V = 0\cdot0858 \times 230$$

$$\underline{V = 19\cdot73 \ \text{V}}$$

(b)   Current in copper cable, $I_1 = \dfrac{V}{R_1}$

(Note that V is the p.d. used to overcome the resistance in the cables.)

$$I_1 = \frac{19\cdot73}{0\cdot138}$$

$$\underline{I_1 = 143 \ \text{A}}$$

Current in aluminium cable, $I_2 = \dfrac{V}{R_2}$

$$I_2 = \frac{19\cdot73}{0\cdot227}$$

$$\underline{I_2 = 87 \ \text{A}}$$

Check: $143 + 87 = 230$ A which is the correct total current.

3.6. Temperature Coefficient of Resistance

3.61 The resistance of most materials is affected by changes in temperature. The majority of metallic conductors increase in resistance when their temperature is raised; for example, the resistance of the tungsten filament of an incandescent lamp increases in resistance when the lamp is switched on and the filament becomes white hot. Materials whose resistance increases when the temperature rises are said to have a positive temperature coefficient of resistance. The resistance of certain alloys such as *Manganin, Constantan* or *Eureka* is almost unaffected by temperature so that wire made from these alloys is used to construct standard resistors. For some materials the resistance falls as the temperature increases; such materials are said to have a negative temperature coefficient of resistance. A good example of a conductor having a negative temperature coefficient of resistance is carbon which, although a good conductor, is not a metal.

3.62 The temperature coefficient of resistance of a material (symbol α) is defined as the increase in resistance of a sample having a resistance of one ohm at 0°C, when its temperature is raised by 1°C. Thus, if a coil having a resistance of one ohm at 0°C is heated to t°C, its resistance becomes $(1 + \alpha t)$ ohms. If a coil has a resistance R_0 ohms at 0°C then when heated to t°C each ohm of the original resistance increases to $(1 + \alpha t)$ ohms and the resistance of the whole coil is given by:

$$R_t = R_0(1 + \alpha t)$$

This relationship can be illustrated by a straight line graph as shown in Fig. 3.9.

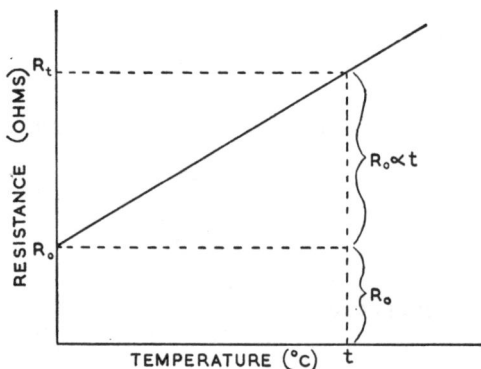

FIGURE 3.9   Increase of Resistance with Temperature.

Example 3.9

A coil has a resistance of 100 ohms at 0°C. If the temperature coefficient of resistance for the coil is 0·004 ohm/ohm°C at 0°C, determine the resistance of the coil at 25°C.

Solution

$$R_t = R_0 (1 + \alpha t)$$
$$R_t = 100 (1 + 0 \cdot 004 \times 25)$$
$$R_t = 100 (1 + 0 \cdot 1)$$
$$R_t = 100 \times 1 \cdot 1$$
$$R_t = 110 \ \Omega$$

3.63 The basic temperature coefficient formula is not very convenient for use in problems where the initial resistance at 0°C is not quoted. A better form of the relationship which can be applied to most problems of this type can be deduced as follows:

Let $\qquad\qquad R_1 =$ the resistance at t_1°C

and $\qquad\qquad R_2 =$ the resistance at t_2°C

Then $\qquad\qquad R_1 = R_0 (1 + \alpha t_1)$ $\qquad\qquad$ ....(i)

and $\qquad\qquad R_2 = R_0 (1 + \alpha t_2)$ $\qquad\qquad$ ....(ii)

Dividing equation (i) by equation (ii)

$$\frac{R_1}{R_2} = \frac{\cancel{R_0}(1 + \alpha t_1)}{\cancel{R_0}(1 + \alpha t_2)}$$

$$\therefore \quad \frac{R_1}{R_2} = \frac{1 + \alpha t_1}{1 + \alpha t_2}$$

Example 3.10

Explain the meaning of "temperature coefficient of resistance". The temperature coefficient of carbon has a negative value, what do you understand by this? A coil of nickel wire of length 85 m and of cross-sectional area 2·8 mm² is connected to a d.c. supply at 25 V. Calculate the current flowing when the temperature of the wire is 110°C. The resistivity of nickel at 0°C is 79×10^{-9} ohm metre. The temperature coefficient of nickel at 0°C is 0·005 ohm per ohm °C rise.

Solution

Temperature coefficient of resistance is a measure of the effect of temperature on the resistance of a material and may be defined as the increase in resistance of a sample of material having a resistance of one ohm at 0°C, when its temperature is increased by 1°C.

The statement "the temperature coefficient of carbon has a negative value" implies that the resistance of carbon falls as its temperature is raised.

Resistance of the coil of nickel wire at 0°C,

$$R_0 = \frac{\rho l}{A}$$

$$R_0 = \frac{79 \times 10^{-9} \times 85}{2 \cdot 8 \times 10^{-6}}$$

$$R_0 = \frac{79 \times 85 \times 1\,000\,000}{1\,000\,000\,000 \times 2 \cdot 8}$$

$$R_0 = 2 \cdot 4 \ \Omega$$

Resistance of coil at 110°C,

$$R_t = R_0(1 + \alpha t)$$
$$R_t = 2 \cdot 4(1 + 0 \cdot 005 \times 110)$$
$$R_t = 2 \cdot 4 \times 1 \cdot 55$$
$$R_t = 3 \cdot 72 \ \Omega$$
$$I = \frac{E}{R}$$
$$I = \frac{25}{3 \cdot 72}$$
$$I = 6 \cdot 72 \ A$$

Example 3.11

The shunt winding of a d.c. generator has a resistance of 150 ohms at a temperature of 15°C. Calculate the resistance when the mean temperature of the winding has reached 45°C. The temperature coefficient of copper is 0·004 28 ohm per ohm °C rise at 0°C.

Solution

$$\frac{R_1}{R_2} = \frac{1 + \alpha t_1}{1 + \alpha t_2}$$

Let R_1 = resistance at 45°C
and R_2 = resistance at 15°C.
Then

$$\frac{R_1}{150} = \frac{1 + (0 \cdot 004 \ 28 \times 45)}{1 + (0 \cdot 004 \ 28 \times 15)}$$

$$\frac{R_1}{150} = \frac{1 \cdot 1926}{1 \cdot 0642}$$

$$\therefore \quad R_1 = 150 \times \frac{1 \cdot 1926}{1 \cdot 0642}$$

$$R_1 = 168 \ \Omega$$

SUMMARY OF FORMULAE

Conductor resistance is directly proportional to the length and inversely proportional to the cross-sectional area of the conductor. Insulation resistance is inversely proportional to the length of cable.

Resistance of conductor: $\quad R = \dfrac{\rho l}{A}$ ohms

Voltage drop equation: $\quad V = E - RI$ volts

Current density:
$$\text{Current Density} = \frac{\text{Current}}{\text{Cross-sectional area}}$$

Temperature coefficient:
$$\begin{cases} R_t = R_0(1 + \alpha t) \text{ ohms} \\ \dfrac{R_1}{R_2} = \dfrac{1 + \alpha t_1}{1 + \alpha t_2} \end{cases}$$

EXERCISE 3

1. Explain with the aid of sketches or diagrams how the insulation resistance of a piece of insulated cable differs from its conductor resistance.
   If a coil of cable 50 m long, has an insulation resistance of 10 000 megohms, and a conductor resistance of 0·275 ohm, calculate the corresponding values for 800 m of the same cable.

2. Calculate the resistance of a single core copper cable 200 m long and 25 mm² in cross-sectional area. Take the resistivity of copper as $17·3 \times 10^{-9} \ \Omega$ m.

3. What do you understand by the term *insulation resistance*, as applied to a length of insulated copper wire lying in a metal conduit?
   Explain with sketches or diagrams how the insulation resistance of the above wire will differ from its conductor resistance.
   A 300 m length of single-core insulated copper wire has an insulation resistance of 15 000 megohms, and a conductor resistance of 1·08 ohms. Calculate the corresponding values for a 75 m length of the same cable.

4. A length of nichrome wire, 1500 mm long and 2 mm diameter is connected to a 5·0 V d.c. supply. Calculate the current in the wire if the resistivity of nichrome is $1110 \times 10^{-9} \ \Omega$ m.

5. A factory is supplied with power at 250 V through a pair of feeders of total resistance 0·025 ohm. The load consists of 200, 100-watt lamps, and four motors, each taking 30 amperes.
   Find: (a) the total current taken.
   (b) the voltage at the station end of the feeders.
   (c) the power lost in the feeders.   U.E.I.

6. A two core cable 1·5 km long has cores of 120 mm² cross-sectional area and is fed at one end at a constant pressure of 250 V. A consumer whose full load current is 15 A is connected to the far end of the cable, and another consumer whose full load current is 45 A is connected to a point 1 km from the feed point. Calculate the p.d. across each load when both consumers are taking full load current. Take the resistivity of the conductors as $17·3 \times 10^{-9} \ \Omega$ m.

7. A d.c. generator supplies a load A of 20 kW at 500 V over a distance of 150 m. At a distance of 100 m from the generator there is also a load B taking a current of 50 A connected to the same cable. If the two core cable has a resistance of 1 Ω per km per core, calculate the voltage at load B and the voltage at the generator. What is the power in kW at B and the total power supplied by the generator?

8. Compare the sizes of an aluminium and a copper conductor designed to carry the same current with the same voltage drop.
   A copper conductor of length 200 m and cross-sectional area 50 mm² is to be replaced by one of aluminium. Calculate the cross-sectional area of the aluminium conductor required, if the resistivity of copper is $17·3 \times 10^{-9} \ \Omega$ m and the resistivity of aluminium is $28·4 \times 10^{-9} \ \Omega$ m.

9. A stepped distribution cable, 180 m long, consists of a 100 m length, *AB*, of 2-core cable of cross-sectional area 70 mm² connected in series with 80 m length, *BC*, of 2-core cable of cross-sectional area 50 mm². The cable is supplied at *A* with a constant pressure of 250 *V*.

   A consumer connected across the cable at *B* takes a full load current of 180 A, and another consumer connected at *C* takes a full load current of 200 A. Find the voltages at each consumer's terminals:
   (a) when neither consumer is taking load,
   (b) when *B* takes full load and *C* takes half full load,
   (c) when both *B* and *C* take full load.
   (The resistivity of copper may be taken as $17 \cdot 3 \times 10^{-9}$ Ω m.)

10. The field winding of a shunt motor has a resistance of 240 ohms at 15°C. If after running on load for some time the resistance of the winding rises to 255 ohms, determine the mean temperature of the winding. The temperature coefficient of resistance for copper may be taken as 0·004 28 ohm per ohm °C at 0°C.

11. Define the term *temperature coefficient of resistance*.
    The resistance at 15°C of a stator winding of a single phase motor is 1·8 Ω. Calculate the temperature of the winding if, after the motor has been running for some time, the resistance is found to be 2·15 Ω. (Assume that the temperature coefficient of resistance of the material of the winding is 0·0042 Ω/Ω °C.)   W.J.E.C.

12. What do you understand by the term temperature coefficient of resistance?
    The temperature coefficient of carbon has a negative value. What does this mean?
    The two ends of a coil of german silver wire 120 m long and of cross-sectional area 2 mm² are connected to a d.c. supply at 85 V. Calculate the current in the coil when the temperature of the wire is 135°C.
    The average temperature coefficient of german silver at 0°C may be taken as 0·000 44 Ω/Ω °C.
    The resistivity of german silver at 0°C may be taken as 410×10^{-9} Ω m.

4 Elementary Mechanics

4.1 Velocity, Speed and Acceleration

4.11 *Speed* is a measure of how quickly an object moves. If a moving object covers equal distances in equal intervals of time it is said to move at a constant speed.

$$\text{Speed} = \frac{\text{Distance moved}}{\text{Time taken}}.$$

As, when using basic SI units, distance is measured in metres and time in seconds, it follows that the SI unit of speed is the metre per second (abbreviation m/s). When considering speeds of vehicles it is often convenient to measure the speed in a non-standard unit such as the kilometre per hour (abbreviation km/h). As 1 km = 1000 m and 1 h = 3600 s, it follows that:

$$1 \text{ km/h} = \frac{1000}{3600} \text{ m/s}.$$

Hence to convert km/h into m/s it is necessary to multiply by $\frac{10}{36}$. The velocity of a moving object is its speed together with information concerning the direction in which it is moving. For example, the speed of an aircraft could be 300 km/h, but its velocity might well be 300 km/h in a South Easterly direction. It may be said that speed is the numerical part or magnitude of a velocity.

Quantities such as velocities which need both a direction as well as a magnitude in order to specify them completely are known as *vector* quantities. Vector quantities, which include velocity, force, and acceleration, can be represented graphically by drawing a line to scale in the appropriate direction. Quantities such as area, volume, age, and cost, where direction is of no significance are known as *scalar* quantities.

4.12 A revolving body has a speed of rotation which is measured by the angle it turns through in a given time. Thus the speed of rotation of a revolving body is given by the equation:

$$\text{Speed of rotation} = \frac{\text{Angle turned through}}{\text{Time taken}}.$$

For theoretical purposes angles are measured in radians, there being 2π radians in a circle, i.e. 2π radians $= 360°$. The speed of rotation as defined above is called the angular velocity (symbol ω); angular velocities are measured in radians per second (abbreviation rad/s). For many practical purposes it is more convenient to measure the rotational speed by determining the number of revolutions completed in a given time. Thus the speed may be measured in revolutions per second (abbreviation rev/s) or, more often, in revolutions per minute (abbreviation rev/min). It is obvious that:

$$\text{60 revolutions per minute} = \text{1 revolution per second.}$$

The symbol N is often used to denote rotational speeds measured in revolutions per minute. To convert angular velocities to speeds in rev/min, and vice versa, it should be noted that 1 revolution $= 2\pi$ radians and so:

$$1 \text{ rad/s} = \frac{1}{2\pi} \text{ rev/s}$$

$$\therefore \quad 1 \text{ rad/s} = \frac{60}{2\pi} \text{ rev/min}$$

or
$$1 \text{ rev/min} = \frac{2\pi}{60} \text{ rad/s}.$$

4.13 When an object moving in a straight line increases its speed it is said to *accelerate*. A decrease in speed is a negative acceleration or deceleration. If the acceleration is uniform then:

$$\text{Acceleration} = \frac{\text{Increase in speed}}{\text{Time taken}}.$$

As speeds are measured in metres per second, it follows that accelerations are measured in metres per second per second (abbreviation m/s²). It is beyond the scope of this text book to discuss the meaning and method of calculating the acceleration of an object which is moving on a curved path, nevertheless it can be said that any change in velocity implies an acceleration, and so accelerations can be caused by changes in the direction of movement as well as by changes in speed. The idea of acceleration can also be applied to rotating bodies; in this case:

$$\text{Angular acceleration} = \frac{\text{Increase in angular velocity}}{\text{Time taken}}.$$

Example 4.1

(a) A car accelerates uniformly along a straight road, starting from rest and attaining a speed of 60 km/h in 50 s. Determine its acceleration in m/s².

(b) The rotor of an electric motor accelerates from 1000 rev/min to 1573 rev/min in 10 s. Determine its angular acceleration in rad/s².

Solution

(a)   Increase in speed $= 60 - 0$

                        Increase in speed $= 60$ km/h.

Converting this to metres per second:

$$\text{Increase in speed} = 60 \times \frac{10}{36}$$

Increase in speed $= 16.7$ m/s.

$$\text{Acceleration} = \frac{\text{Increase in speed}}{\text{Time taken}}$$

$$\text{Acceleration} = \frac{16.7}{50}$$

Acceleration $= 0.334$ m/s².

(b)   Increase in speed $= 1573 - 1000$

                        Increase in speed $= 573$ rev/min.

Converting this to radians per second:

$$\text{Increase in speed} = 573 \times \frac{2\pi}{60}$$

Increase in speed $= 60$ rad/s.

$$\text{Acceleration} = \frac{\text{Increase in speed}}{\text{Time taken}}$$

$$\text{Acceleration} = \frac{60}{10}$$

Acceleration $= 6$ rad/s².

4.2 Force, Mass and Weight

4.21 A *force* (symbol F) can be described as a push or a pull. When a force acts on a body it may:

(a) accelerate the body, i.e. increase its speed or change the direction in which it moves,
(b) decelerate the body, i.e. reduce its speed,
(c) deform the body, i.e. change its shape,
(d) be exactly resisted by other forces.

As the action of a force is always to attempt to alter the way in which a body is moving, the following statement provides a good definition of a force. A force is that which changes or tends to change a body's state of rest or its uniform motion in a straight line. When using SI units, forces are measured in **newtons** (abbreviation N). Note that a force needs both magnitude and direction to define it and is thus a vector quantity.

4.22 The *mass* (symbol *m*) of a body is the amount of substance contained in it. Mass is measured in **kilogrammes** (abbreviation kg). The mass of a body may be determined by weighing it: the *weight* of a body is actually the force exerted on it by the action of gravity. It follows that the weight of a body should strictly be measured as a force in newtons even though spring balances and other types of equipment used for weighing are often calibrated to read the mass of the body in kilogrammes. It is possible to calibrate spring balances in kilogrammes only because the strength of the earth's gravitational field varies but slightly from place to place on the earth's surface. It is found that a mass of 1 kg weighs 9·81 newtons when acted on by standard earth gravity (see Section 4.23 below). Thus if the scale mark on the spring balance which actually indicates 9·81 newtons is labelled as 1 kg, the spring balance reading can be taken as indicating the mass if used under earth conditions. Only for very accurate work would the small variations in earth's gravity have to be allowed for. If the strength of the gravitational field alters, the weight alters even though the mass does not, and so, if a mass of 1 kg were taken to the moon and weighed on a spring balance which was calibrated to read mass on earth, the balance would indicate only 0·16 kg, i.e. the balance calibration would be inaccurate for the purpose of determining mass on the moon. Nevertheless the weight of the mass in newtons would still be the balance reading multiplied by 9·81, i.e. the weight would have only 0·16 of its value on earth, even though the mass is unchanged. It follows that a spring balance calibrated in newtons always indicates forces or weights correctly, but a spring balance calibrated in kilogrammes can be taken as indicating mass only when used on the earth, although the true force or weight in newtons can always be found by multiplying its reading by 9·81.

4.23 It has been stated that one effect of applying a force to a body is to cause it to accelerate or increase speed. Sir Isaac Newton showed that there is an exact relation between the mass of a body, the force applied to it and the resulting acceleration. This relation is expressed by the formula:

$$F = m \times a.$$

If a body is allowed to fall freely in a gravitational field the force acting on it is its own weight, thus if the acceleration due to gravity is *g* it follows that:

weight (F) = mass $(m) \times g$ (acceleration due to gravity).

Using SI units, the acceleration due to the earth's gravitational field is:

$$g = 9·81 \text{ metres per second per second.}$$

It follows that:

weight (in newtons) = mass (in kg) × 9·81 (m/s²)

for a mass on the earth's surface.

4.3 Springs

4.31 If a helical spring is stretched by applying a force to it, it is found that the extension of the spring is directly proportional to the force applied, provided that the coils of the spring are not pulled too far apart.

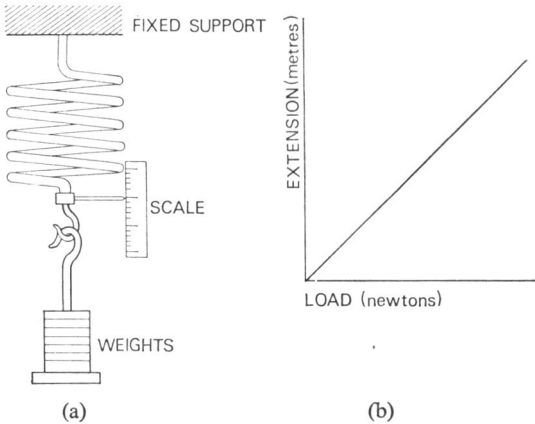

FIGURE 4.1 Extension of a Helical Spring.
(a) Spring being tested.
(b) Extension-load graph.

Figure 4.1(a) shows a helical spring being tested by loading it with weights, and Fig. 4.1(b) shows a typical graph of the results obtained. It can be seen that the graph is linear. Thus a force can be measured by noting how much it extends a spring which is the principle used by the spring balance. Figure 4.2 shows a typical spring balance.

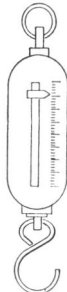

FIGURE 4.2 Spring Balance.

4.32 The *rate* or *stiffness* of a spring is the value of force which extends it by one unit of length and so:

$$\text{Force} = \text{Extension} \times \text{rate of spring}.$$

This formula applies whenever the extension-load graph is linear. For a helical spring this means that the coils are neither pulled too far apart, nor compressed so closely together that they touch each other.

Example 4.2

A spring is found to extend by 20 mm when a load of 40 N is applied to it. (a) What is the rate of the spring? (b) If a load of 60 N were applied, by how much would the spring extend?

Solution

(a)
$$\text{Force} = \text{Extension} \times \text{rate of spring}$$

$$\therefore \quad \text{Rate of spring} = \frac{\text{Force}}{\text{Extension}}$$

$$\text{Rate of spring} = \frac{40}{20 \times 10^{-3}}$$

(Note that the factor 10^{-3} is required to convert the extension in millimetres into a value in metres)

$$\underline{\text{Rate of spring} = 2 \times 10^3 \text{ N/m.}}$$

(b)
$$\text{Extension} = \frac{\text{Force}}{\text{Rate of spring}}$$

$$\text{Extension} = \frac{60}{2 \times 10^3}$$

$$\underline{\text{Extension} = 30 \times 10^{-3} \text{ m} \quad \text{or} \quad 30 \text{ mm}}$$

4.33 Another form of spring is the *spiral* spring, which is shown in Fig. 4.3. This type of spring may be used to control the rotation of a spindle, for example in an electric meter. If a twisting force or *torque*

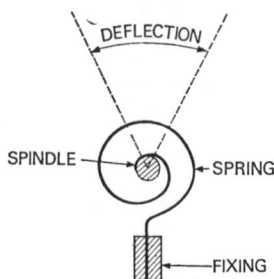

FIGURE 4.3 Spiral Spring.

(see Section 4.5) is applied so as to cause the spindle to rotate, the angle through which the shaft rotates is directly proportional to the torque applied. The rate of such a spring is the amount of torque (measured in N m) required to rotate the shaft by a unit angle (one radian), and so:

Torque (in N m) = Deflection (in radians) × rate of spring (in N m/rad).

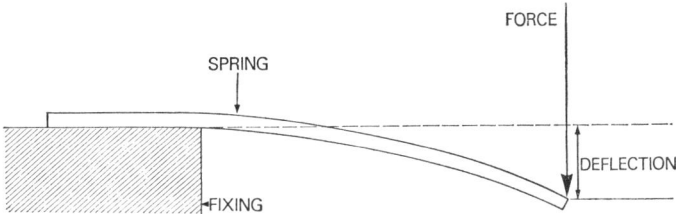

FIGURE 4.4   Cantilever Spring.

4.34   Figure 4.4 shows a cantilever spring which is deflected by applying a force to its end. The methods of calculating accurately the deflection of this type of spring are complicated, and beyond the scope of this text book. Nevertheless, if the deflection is small compared with the length of the spring, simplification is possible and it is found that once again the deflection produced by a force is directly proportional to the force.

Example 4.3

A cantilever spring 500 mm long, is found to deflect by 20 mm when a force of 50 N is applied to its end. What deflection is produced by a force of 30 N?

Solution

As deflection is proportional to force:

$$\frac{\text{Final force}}{\text{Final deflection}} = \frac{\text{Initial force}}{\text{Initial deflection}}$$

As this formula is a relation between two similar ratios it is independent of the units used, provided that the same units are used on each side of the equation, so in this case it is not necessary to convert the deflections given in millimetres into metres.

Thus
$$\frac{30}{\text{Final deflection}} = \frac{50}{20}$$

or
$$\text{Final deflection} = \frac{30 \times 20}{50}$$

$$\text{Final deflection} = 12 \text{ mm}$$

61

4.4 Work and Energy

4.41 *Work* (symbol *W*) is done whenever an object is moved by applying a force to it. The work done is given by the product of the force acting and the distance which the body moves *in the direction of the force.*

Thus $$W = F \times d.$$

As force is measured in newtons and distance in metres, the unit of work is the newton metre (N m) which is also called the **joule** (abbreviation J).

Example 4.4

What work is done in lifting a mass of 10 kg through a height of 5 m?

Solution

Force required to lift 10 kg:

$$\text{Weight, } F = mg$$
$$F = 10 \times 9\cdot81$$
$$F = 98\cdot1 \text{ N}$$
$$W = F \times d$$
$$W = 98\cdot1 \times 5$$
$$W = 490\cdot5 \text{ J}$$

4.42 *Energy* is the ability to do work. There are many forms of energy, some of the more important forms being mechanical energy, heat energy, chemical energy, electrical energy, and atomic energy. Any form of energy can be converted into another by using suitable equipment. For example, consider a lump of coal fed into the furnace of a power station boiler. The coal possesses chemical energy and, as it burns, this energy is converted into heat energy. The heat is absorbed by the water in the boiler, forming steam, and this drives a turbine which produces mechanical

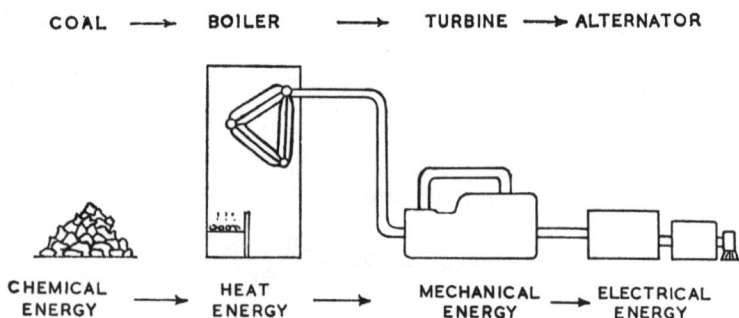

COAL ⟶ BOILER ⟶ TURBINE ⟶ ALTERNATOR

CHEMICAL ENERGY ⟶ HEAT ENERGY ⟶ MECHANICAL ENERGY ⟶ ELECTRICAL ENERGY

FIGURE 4.5 Energy Conversion.

energy. The turbine, in turn, drives the generator which produces the output of the power station in the form of electrical energy.

4.43 The law of *conservation of energy* states that energy can neither be created nor destroyed. So in any energy conversion process, the total energy coming out of the process is equal to the total amount of energy put in. One advantage of using the SI system of units is that all forms of energy are normally measured using the same unit, that is the joule. However it is often convenient in electrical work to measure energy in kW h. It should be remembered that to convert kW h into joules, multiply by 3 600 000.

4.44 Although as is stated in Section 4.43, all the energy put into a process produces an equivalent output, the output is very seldom all in the required form. For example an electric motor is required to convert electrical energy into mechanical energy, but it also produces heat and noise which are not required. Some of the energy input is inevitably converted into unwanted forms and so is wasted. It follows that the useful energy output of a machine is normally less than the input, the difference being due to energy being converted into unwanted forms and so wasted. The *efficiency* of a machine is the ratio between the useful output and the input, and is often expressed as a percentage.

$$\text{Efficiency}\% = \frac{\text{Useful energy output}}{\text{Total energy input}} \times 100$$

When using the above formula it is important to remember that the output and input must be measured in the same units. Note too that the efficiency of even the best machine is always less than 100%.

Example 4.5

Calculate the energy in kW h required to accomplish each of the following tasks.

(a) To raise 200 litres of water through a head of 50 m, if the efficiency of the process is 50% (1 litre of water has a mass of 1 kg).

(b) To lift a mass of 250 kg through 300 metres, if the efficiency of the lifting gear is 30%.

Solution

(a) Force required to raise water $= 200 \times 9 \cdot 81$ N (as 1 kg weighs $9 \cdot 81$ N)

$$F = 1962 \text{ N}$$
$$W = F \times d$$

$\therefore$ Useful work done in raising water $= 1962 \times 50$

$$\text{Useful work} = 98\ 100 \text{ J}$$

But the efficiency is only 50%.

$$\therefore \quad \text{Total work done} = 98\,100 \times \frac{100}{50}\,\text{J}$$

.e. $\qquad$ Total work $= 196\,200\,\text{J}$

This is the amount of energy which must be put into the operation,

and since $1\,\text{kW h} = 3\,600\,000\,\text{joule}$

$$\text{Energy required} = \frac{196\,200}{3\,600\,000}$$

$$\underline{\text{Energy required} = 0\cdot0545\,\text{kW h}}$$

(b) Force required to lift the mass $= 250 \times 9\cdot81\,\text{N}$
$$F = 2453\,\text{N}$$

Useful work done $= 2453 \times 300\,\text{joule}$

Useful work done $= 735\,900\,\text{joule}$

$$\text{Total work done} = 735\,900 \times \frac{100}{30}$$

Total work done $= 2\,453\,000\,\text{joule}$

$$\therefore \quad \text{Energy required in kW h} = \frac{2\,453\,000}{3\,600\,000}$$

$$\underline{\text{Energy} = 0\cdot68\,\text{kW h}}$$

4.5 Power

4.51 *Power* (symbol P) is a measure of how quickly work is done. Thus $P = \text{Time rate of doing work} = \dfrac{\text{Work done}}{\text{Time taken}}$. The SI unit of power is the **watt** (abbreviation W) one watt being equivalent to work being done at the rate of one joule per second;

$\qquad$ one watt $=$ one joule per second
or $\qquad$ one watt $=$ one newton metre per second.

It is very convenient that the watt is also the unit used to measure electrical power.

Example 4.6

A hoist raises a load of 3 t through 8 m in 15 seconds. Determine the power developed by the hoist. (1 t = 1000 kg).

Solution

$$\text{Force required to lift the load} = 3 \times 1000 \times 9{\cdot}81$$
$$\text{Force} = 29\,430\,\text{N}$$
$$\text{Work done by hoist} = \text{force} \times \text{distance}$$
$$\text{Work done} = 29\,430 \times 8$$
$$\text{Work done} = 235\,440\,\text{J}$$

$$\text{Power} = \text{rate of doing work} = \frac{\text{work done}}{\text{time taken}}$$

$$\text{Power} = \frac{235\,000}{15}$$

$$\text{Power} = 15\,696\,\text{W or } 15{\cdot}7\,\text{kW}$$

4.52   As the useful energy output of a process is always in practice less than the energy input, it follows that the power output of a machine is always less than the power input. The efficiency of a machine was given in section 4.44 as:

$$\text{Efficiency}\% = \frac{\text{Useful energy output}}{\text{Total energy input}} \times 100$$

Consider the machine acting for just one second, then:

$$\text{Efficiency}\% = \frac{\text{Useful energy output in one second}}{\text{Total energy input in one second}} \times 100$$

But the amount of energy used, or work done, in one second is the power, thus:

$$\text{Efficiency}\% = \frac{\text{Useful power output}}{\text{Total power input}} \times 100$$

Example 4.7

A 460 volt d.c. motor is used to drive a rotary pump. The pump raises water from a well 50 m deep at the rate of 90 000 litre per hour. The efficiency of the motor is 85 per cent and of the pump and pipe system is 75 per cent.
Calculate the current supplied to the motor.

Solution

When solving a problem of this type it is important to keep clearly in mind which powers constitute the inputs and outputs of the two machines involved. A flow diagram, as in Fig. 4.6, is a great help in this respect.

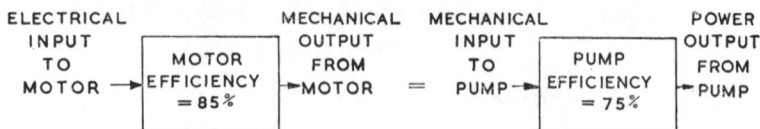

FIGURE 4.6   Flow Diagram for Example 4.7.

$$\text{Work done by pump in one second} = \frac{90\ 000 \times 9{\cdot}81 \times 50}{3600}$$

$$= 12\ 262 \text{ J/s}$$

$$\text{Power output of pump} = 12{\cdot}26 \text{ kW}$$

Allowing for the pump efficiency:

$$\text{Power input to pump} = 12{\cdot}26 \times \frac{100}{75}$$

$$P = 16{\cdot}35 \text{ kW}.$$

By referring to the flow diagram (Fig. 4.3) it can be seen that this is also the power output of the motor and, therefore, allowing for the motor efficiency:

$$\text{Power input to motor} = 16{\cdot}35 \times \frac{100}{85}$$

$$P = 19{\cdot}24 \text{ kW}.$$

$$I = \frac{P}{E}$$

$$I = \frac{19\ 240}{460}$$

$$\underline{I = 41.8 \text{ A}}$$

4.6 Moment and Torque

4.61   Figure 4.4 shows two cases in which a bar hinged at A is acted on by a force at point B. In each case the force will tend to make the bar rotate about A. The magnitude of this turning effect, which is called the *moment* of the force about A, will depend both on the magnitude of the force and on its distance from A. The moment of a force is equal to the force multiplied by its distance from the hinge or pivot; this distance

66

TURNING EFFECT

(a)

MOMENT = F x d

(b)

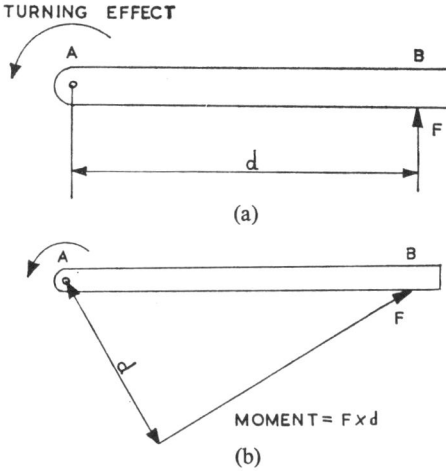

FIGURE 4.7 Moment of Force.
(a) Force Acting at Right Angles to Bar.
(b) Force Not at Right Angles to the Bar.

must be measured at right angles to the line of action of the force as shown in Fig. 4.7(b).

$$\text{Moment} = F \times d$$

As force is measured in Newtons and distance in metres, it follows that the moment is measured in Newton metres (abbreviation N m).

4.62 The turning effect exerted by a shaft, for example the shaft of an electric motor, is called the *torque*. Torque is measured in the same units as moment i.e. in N m. Consider a shaft driving a pulley as shown in Fig. 4.8. If the pulley is raising a load at a constant speed then the torque

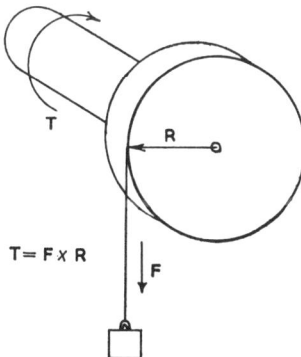

T = F x R

FIGURE 4.8   Torque Required to Raise a Load.

in the shaft must equal the moment caused by the force required to lift
the load acting at the circumference of the pulley. The force required to
lift the load is measured in newtons, and the pulley radius in metres. In
this case if the shaft makes N revolutions per minute then the load will
be raised $2\pi RN$ metres in every minute. Work will be done at the rate of
$2\pi RNF$ joule/min. But $F \times R$ equals the torque, T, and so:

$$\text{Rate of doing work} = 2\pi NT \text{ joule/min.}$$

$$\text{Power} = \frac{2\pi NT}{60} \text{ joule/second or watts}$$

4.63 Motors are often tested by using a brake, the torque developed
during the test being determined by measuring the force required to

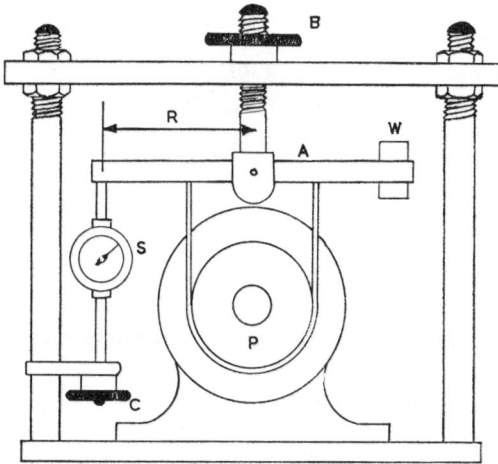

FIGURE 4.9   Simple Brake for Motor Testing.

restrain the brake. Figure 4.9 shows a simple but effective type of brake
suitable for testing small electric motors. The brake consists of a pivoted
bar, A, which supports a brake band of friction material, F, often a
webbing belt. The arm can be raised or lowered by means of an adjusting
nut B. When the arm is lowered so that the brake band is not in contact
with the motor pulley P, the pivoted bar is balanced by the weight W,
so that a zero reading is indicated on the spring balance S. When the bar
is raised, by tightening nut B, the brake band makes contact with the pulley,
so loading the motor, the torque developed being given by:

$$\text{Torque} = \text{Spring balance reading} \times \text{brake arm radius } (R)$$

The load on the motor can be increased to any required value by tightening
nut B. The second adjusting nut C is provided so that the bar can be set
to a horizontal position before taking any reading, as this is essential to

68

ensure accurate results. N.B. When calculating the moment required to restrain the brake, the distance R must be measured at right angles to the restraining force due to the spring balance.

4.64 The *rating* of a motor is the maximum output power that it is intended to produce; if a motor is used to drive a load requiring more power than it is designed to produce, the motor may well overheat and burn out. On the other hand a motor can be used to produce less than its rated power although some loss of efficiency results when a motor is used to drive loads requiring only a small fraction of its rated output.

Example 4.8

The following results were obtained during a test on a d.c. motor:

Brake load	27 N
Effective brake arm radius	0·5 m
Motor speed	1100 rev/min
Supply pressure	240 V
Motor current	8 A

Calculate the power output, power input and efficiency of the motor for the conditions of the test.

Solution

$$\text{Torque} = Fr$$

$$T = 27 \times 0\cdot5$$

$$T = 13\cdot5 \text{ N m}$$

$$\text{Power output} = \frac{2\pi NT}{60}$$

$$P = \frac{2\pi \times 1100 \times 13\cdot5}{60}$$

$$\underline{P = 1554 \text{ W}}$$

$$\text{Power input} = VI$$

$$P = 240 \times 8$$

$$\underline{P = 1920 \text{ W}}$$

$$\text{Efficiency}\% = \frac{\text{Output}}{\text{Input}} \times 100$$

$$\text{Efficiency}\% = \frac{1554}{1920} \times 100$$

$$\underline{\text{Efficiency} = 80\cdot9\%}$$

ELECTRICAL PRINCIPLES

4.7 Machines

4.71 A machine may be described as a device which accepts an energy input in one form and produces an energy output in some other form. There are many simple machines in which the input or effort and the output or load are both in the form of mechanical energy, the machine acting so as to change the direction or alter the value of the effort force applied. Examples of this type of machine include pulley systems, winches, levers, screwjacks etc. There are two important ratios which are applied to such machines;

$$\text{Velocity Ratio (abbreviation V.R.)} = \frac{\text{Speed of effort}}{\text{Speed of load}}$$

and

$$\text{Mechanical Advantage (abbreviation M.A.)} = \frac{\text{Force acting on load}}{\text{Force applied as effort}}$$

It should be noted that as both load and effort move at the same time the expression given above for the velocity ratio also implies that:

$$\text{V.R.} = \frac{\text{Distance moved by effort}}{\text{Distance moved by load}}$$

4.72 If a machine is perfect, i.e. its efficiency is 100%, then the work done at the input equals the work done at the output. In this event we have: Force applied as effort × Distance moved by effort = Force acting on load × Distance moved by load.

$$\text{Hence:} \quad \frac{\text{Distance moved by effort}}{\text{Distance moved by load}} = \frac{\text{Force acting on load}}{\text{Force applied as effort}}$$

$$\text{i.e.} \quad \text{V.R.} = \text{M.A.}$$

For a practical machine the velocity ratio is usually fixed by the construction of the machine, but the mechanical advantage is less than the theoretical value because of the adverse effects of friction. In this case:

$$\text{Efficiency}\% = \frac{\text{M.A.}}{\text{V.R.}} \times 100$$

4.73 An example of a simple machine is the lever. Figure 4.10 shows several examples of levers. The point about which the lever pivots is called its *fulcrum* and in every case the velocity ratio of the lever is given by:

$$\text{V.R.} = \frac{\text{Distance of effort from fulcrum}}{\text{Distance of load from fulcrum}}$$

In some cases one lever acts upon another thus giving a *compound* lever arrangement. One example of this type of arrangement is shown in Fig. 4.11, another in Fig. 4.12. Consider the system shown in Fig. 4.11. The
70

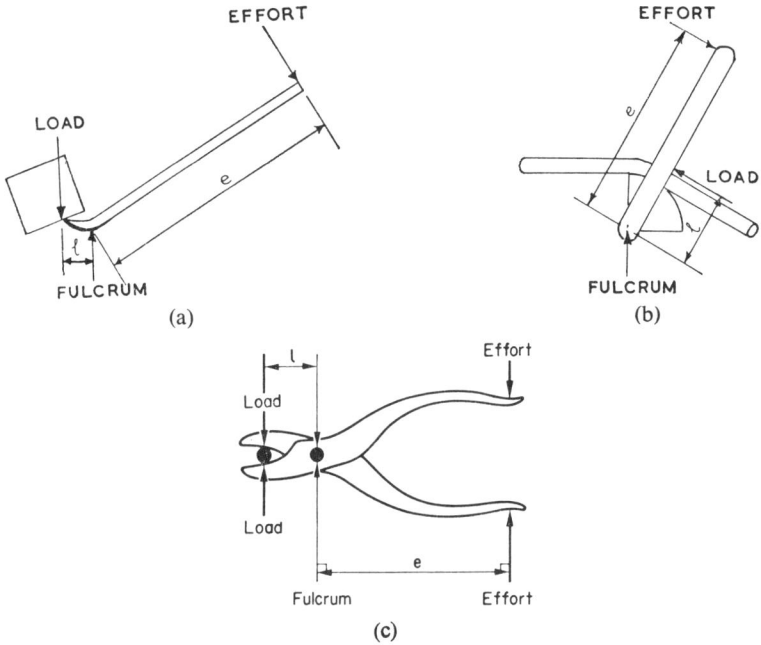

FIGURE 4.10   Examples of Levers.
(a) Crow bar.
(b) Pipe bending machine.
(c) Side cutters.

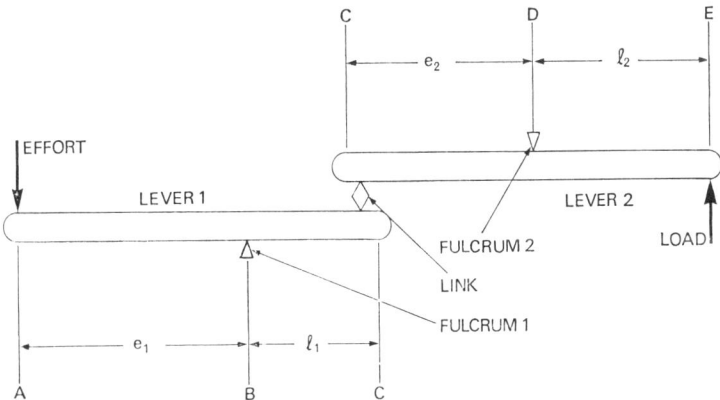

FIGURE 4.11   Compound Lever.

71

effort applied at A moves A downwards and C upwards. The velocity ratio of lever 1 is $V.R._1 = \dfrac{e_1}{l_1}$ hence the distance moved by C is given by:

$$\text{Distance moved by } C = \text{Distance moved by } A \times \frac{l_1}{e_1}$$

Distance moved by C = Distance moved by $A \times 1/V.R._1$

But the levers are linked at C, so the end C of lever 2 is moved upwards while E moves downwards. The velocity ratio of lever 2 is:

$$V.R._2 = \frac{e_2}{l_2}$$

It follows that:

Distance moved by E = Distance moved by $C \times 1/V.R._2$

Distance moved by E = Distance moved by $A \times 1/V.R._1 \times 1/V.R._2$

The overall velocity ratio is $\dfrac{\text{distance moved by } A}{\text{distance moved by } E}$

$$\text{Overall V.R.} = \frac{\text{Distance moved by } A}{\text{Distance moved by } A \times 1/V.R._1 \times 1/V.R._2}$$

Overall V.R. = $V.R._1 \times V.R._2$

This result can be extended to any number of levers, i.e:

$$\text{Overall V.R.} = V.R._1 \times V.R._2 \times V.R._3 \times \text{etc.}$$

Example 4.9

Figure 4.12 shows the arrangement of a compound lever used in a testing machine. Assuming an efficiency of 90%, determine the force

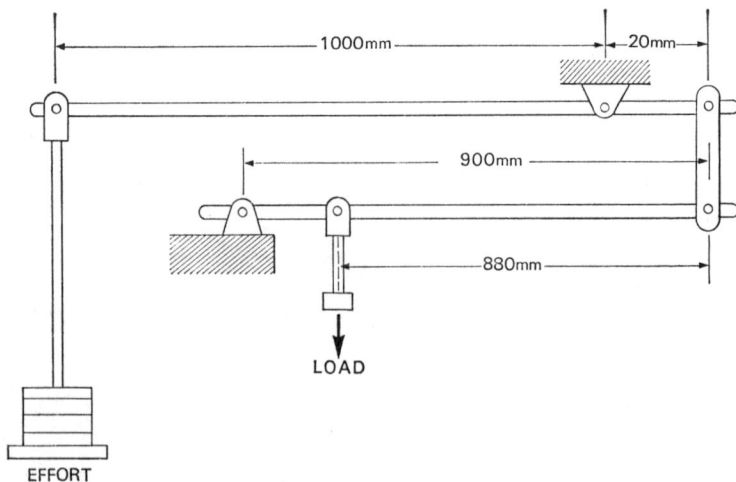

FIGURE 4.12   Example 4.9.

applied to the load when the effort is produced by a mass of 20 kg. (1 kg weights 9·81 N.)

Solution

For first lever:

$$V.R._1 = \frac{1000}{20} = 50$$

Note that the distances from the fulcrum must each be expressed in the same units. In the calculation above each has been expressed in millimetres.

For second lever:

The distance of the load from the fulcrum is $900 - 880 = 20$ mm and so:

$$V.R._2 = \frac{900}{20} = 45$$

$$\text{Overall V.R.} = V.R._1 \times V.R._2$$

$$\text{Overall V.R.} = 50 \times 45$$

$$\text{Overall V.R.} = 2250$$

$$\text{M.A.} = V.R. \times \text{efficiency}$$

$$\text{M.A.} = 2250 \times \frac{90}{100}$$

$$\text{M.A.} = 2025$$

$$\text{Force applied to load} = \text{Effort} \times \text{M.A.}$$

$$F = 20 \times 9·81 \times 2\,025$$

$$\underline{F = 307\,305 \text{ N or } 307·3 \text{ kN (correct to 4 figures)}}$$

It can be seen that a compound lever arrangement is capable of providing a great magnification of the effort force.

4.74 Figure 4.13 shows a *toggle* comprising two links *A* and *B*. If a

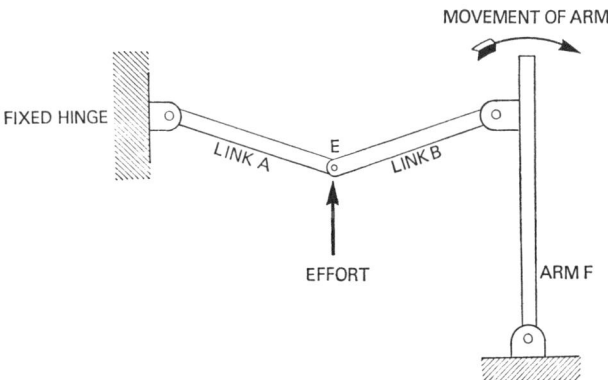

FIGURE 4.13   Toggle.

force is applied to the hinge at E, this causes the arm F to move in the direction shown. Toggle arrangements are often used in switches, for they are capable of giving a high mechanical advantage, which increases rapidly as the two parts of the toggle come more in line with each other. An example, showing how the force exerted by a toggle can be determined using the principle of the triangle of forces is given in section 4.113.

4.75 Another device which is capable of providing a high mechanical advantage is the cam. Figure 4.14 shows a circular cam, which is simply a circular piece of metal fitted eccentrically to a shaft. As the shaft revolves the cam follower, which is kept in contact with the cam profile by means of a spring, is alternately raised and allowed to fall. Cams need not be

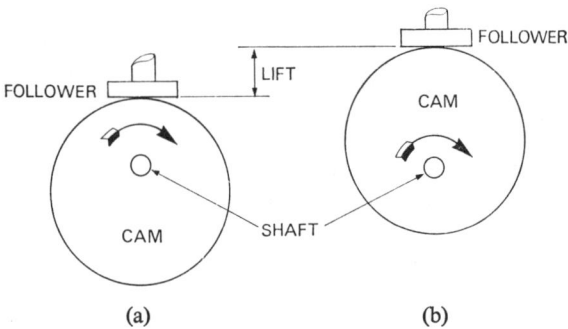

FIGURE 4.14   Cam.
(a) Cam with follower in lowest position.
(b) The same cam after rotating 180° so that follower is in highest position.

circular, but the M.A. obtained depends on the shape of the cam and on the amount of lift it provides to the cam follower. Both toggles and cams are extremely useful devices in switchgear. For example they may be designed to hold contacts in position with very little effort, so that although a strong closing force is applied to the contacts, only a small force applied to the toggle or cam is needed to release the contacts. This force can easily be provided by some form of automatic protective device so giving automatic circuit breaker operation.

4.8   Pulleys

4.81 Another simple type of machine is the pulley, some simple pulley arrangements being shown in Fig. 4.15. The velocity ratio of a simple pulley arrangement is equal to the number of active ropes between the sheaves. Consider for example the situation illustrated in Fig. 4.15(c), where there are 4 ropes between the sheaves. If the load is to be raised by 1 m then each of these ropes must shorten by 1 m and the effort rope must

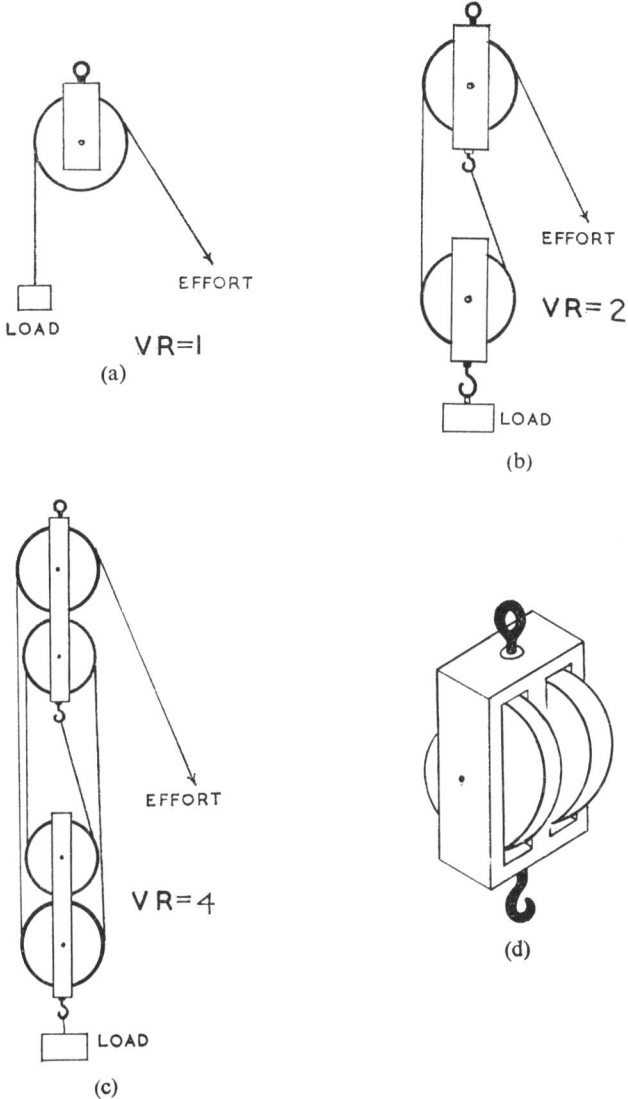

FIGURE 4.15   Pulley (Block and Tackle) Arrangements.
(a) Single pulley used to change direction of effort.
(b) Two pulleys arranged to give V.R. = 2
(c) Four pulleys arranged to give V.R. = 4
(d) Pulley sheave

It should be noted that although the pulleys in diagrams (a), (b) and (c) are shown in line, the most common practical arrangement is to mount the pulleys side by side in a pulley sheave as shown in (d).

be pulled through a distance of 4 m in order to do this. Thus:

$$\text{V.R.} = \frac{\text{Distance moved by effort}}{\text{Distance moved by load}}$$

$$\text{V.R.} = \frac{4}{1} = 4$$

Example 4.10

A set of pulley blocks, having 4 pulleys in the top block and 3 pulleys in the lower block, is fixed to a ceiling beam in a workshop. It is to be used to lift a motor weighing 6300 N from its bedplate. The efficiency of the tackle is 60 per cent.

Calculate the pull required on the free end of the rope to raise the motor. Make a diagrammatic sketch of the tackle in use.

Solution

Figure 4.16 shows the diagrammatic sketch of the tackle. As there are 7 ropes between the pulleys the velocity ratio of the tackle is 7.

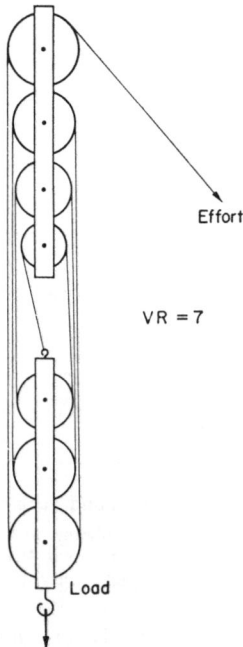

FIGURE 4.16   Arrangement of Tackle Example 4.6.

M.A. $=$ V.R. $\times$ efficiency

M.A. $= 7 \times \dfrac{60}{100}$

M.A. $= 4 \cdot 2$

Effort $= \dfrac{\text{Load}}{\text{M.A.}}$

Effort $= \dfrac{6300}{4 \cdot 2}$

Effort $= 1500\,\text{N}$

4.82 A motor is often coupled to its load using a belt and pulley arrangement as shown in Fig. 4.17. Such an arrangement not only provides a convenient coupling arrangement but also can be arranged to give a

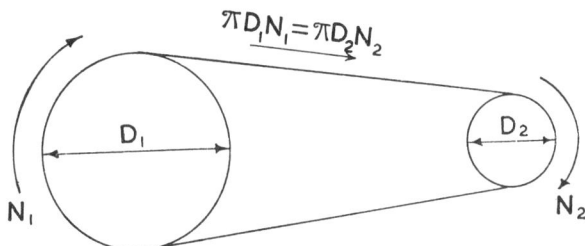

$$\pi D_1 N_1 = \pi D_2 N_2$$

$$\frac{N_2}{N_1} = \frac{D_1}{D_2}$$

FIGURE 4.17   Belt Drive.

speed change. Consider a pulley having a diameter of D metres, running at N rev/min. A spot on the circumference must travel once completely around the circumference in every revolution, that is it travels a distance of πD metres. The speed of the spot must therefore be πDN metres/min or $\dfrac{\pi DN}{60}$ metres/second, and this is also the linear speed of the belt around the pulley provided it does not slip.

Speed of belt $= \pi \times$ diameter of pulley $\times$ speed of pulley.

When the belt links two pulleys, as in Fig. 4.17 the belt speed is the same, whichever of the two pulleys is considered. Thus, if the first pulley has diameter D_1 and speed N_1, and the second pulley has diameter D_2 and speed N_2 then:

$$D_1 N_1 = D_2 N_2$$

$$\frac{N_1}{N_2} = \frac{D_2}{D_1}$$

This result is often expressed in words as:

$$\frac{\text{Speed of driven pulley}}{\text{Speed of driver pulley}} = \frac{\text{Diameter of driver pulley}}{\text{Diameter of driven pulley}}$$

It should be noted that the larger the pulley, the slower is its relative speed. The work done by the driver pulley equals the work done on the driven pulley and so

$$\frac{\text{Torque of driver}}{\text{Torque of driven}} = \frac{\text{Diameter of driver}}{\text{Diameter of driven}}$$

i.e. $\dfrac{N_2}{N_1} = \dfrac{D_1}{D_2} = \dfrac{T_1}{T_2}$

Example 4.11

A drill is driven by a vee-belt from a small electric motor. The motor runs at 1410 rev/min and is fitted with a 50 mm diameter pulley. The drill is to run at 470 rev/min and requires a torque of 2·7 N m. Assuming no belt slip determine (a) the diameter of the drill pulley (b) the speed of the belt (c) the motor torque (d) the power output of the motor.

Solution

(a) $\dfrac{\text{Speed of driven pulley}}{\text{Speed of driver pulley}} = \dfrac{\text{Diameter of driver pulley}}{\text{Diameter of driven pulley}}$

$$\frac{470}{1410} = \frac{50}{\text{Diameter of drill pulley}}$$

$$\text{Diameter of drill pulley} = \frac{5 \times 1410}{470}$$

$$\text{Diameter of drill pulley} = 150 \text{ mm}$$

(b) Belt speed $= \pi \times$ diameter of pulley $\times$ speed of pulley

$$\text{Belt speed} = \frac{\pi \times 50 \times 1410}{60}$$

Belt speed $= 3690$ mm/s, or 3·69 m/s

(c) $\dfrac{\text{Torque of driver}}{\text{Torque of driven}} = \dfrac{\text{Diameter of driver}}{\text{Diameter of driven}}$

$$\frac{\text{Torque of motor}}{2 \cdot 7} = \frac{50}{150}$$

$$\text{Torque of motor} = \frac{2 \cdot 7 \times 50}{150}$$

$$\text{Torque of motor} = 0 \cdot 9 \text{ N m}$$

(d)
$$\text{Power} = \frac{2\pi NT}{60} \text{ watts}$$

$$\text{Power} = \frac{2\pi \times 1410 \times 0.9}{60}$$

$$\underline{\text{Power} = 133 \text{ W}}$$

4.9 Screw Jack

4.91 Another type of machine is the screw-jack, shown in Fig. 4.18. When the handle of the screwjack is turned through one revolution the

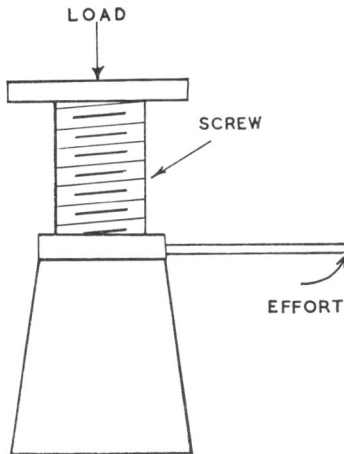

FIGURE 4.18 Simple Screwjack.

load is raised by an amount equal to the pitch of the screw thread; thus the velocity ratio of the simple screwjack is:

$$\text{V.R.} = \frac{2\pi \times \text{length of handle}}{\text{pitch of jack thread}}$$

The velocity ratio of a screwjack can be made very high and so, even though the efficiency is often low, the mechanical advantage is high, and a heavy load may be raised by applying a comparatively small effort.

Example 4.12

A simple screwjack has a thread of 500 threads per metre. The jack handle is 300 mm long, and the efficiency of the jack is 20%. What value of effort applied to the end of the jack handle is required to raise a load weighing 20 kN?

Solution

$$\text{Pitch of screw thread} = \frac{1000}{500}$$

$$\text{Pitch} = 2 \text{ mm.}$$

i.e. the load is raised 2 mm by one revolution of the jack handle.

$$\text{V.R.} = \frac{2\pi \times 300}{2}$$

Note that the length of the handle and the pitch of the screw must both be expressed in the same units, in this case both are in millimetres.

$$\text{V.R.} = 942 \cdot 6$$

$$\text{M.A.} = \text{V.R.} \times \text{efficiency}$$

$$\text{M.A.} = 942 \cdot 6 \times \frac{20}{100}$$

$$\text{M.A.} = 188 \cdot 5$$

$$\text{Effort} = \frac{\text{Load}}{\text{M.A.}}$$

$$\text{Effort} = \frac{20\ 000}{188 \cdot 5}$$

$$\underline{\text{Effort} = 106 \text{ N}}$$

4.10 Winches and Gears

4.101 Loads may be raised by using a winch. Figure 4.19(a) shows a simple winch, which consists of a drum around which a rope is wound. The drum can be rotated by means of a crank handle so winding up or lowering the load. The velocity ratio of a simple winch can be determined by considering it to be a simple lever. It can be seen from the simplified end view shown in Fig. 4.19(b) that the axle of the drum provides the fulcrum, the radius of the crank gives the distance of the effort from the fulcrum and the radius of the drum gives the distance of the load from the fulcrum. It follows that for a simple winch

$$\text{V.R.} = \frac{\text{Radius of crank}}{\text{Radius of drum}}$$

4.102 The simple winch suffers from the disadvantage that, if the crank handle is released, it can spin freely around so allowing the load to fall. This can be prevented by fitting the drum with a ratchet wheel as shown in Fig. 4.20. The teeth on the ratchet wheel are so shaped that,

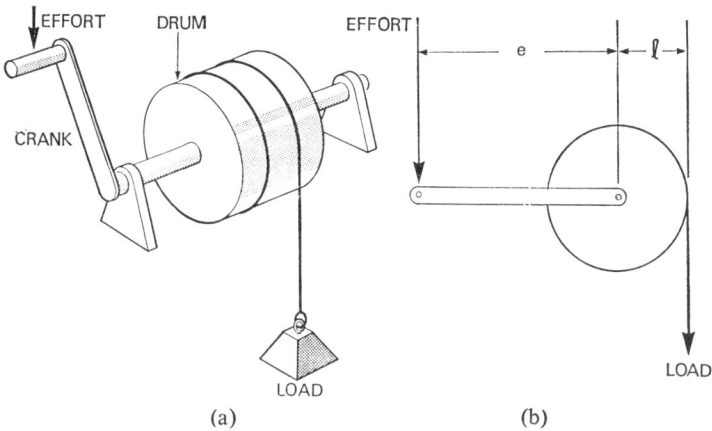

FIGURE 4.19 Simple Winch.
(a) Crank and drum.
(b) End view of winch.

when the drum is turned in the direction which raises the load, the catch
or pawl lifts and clicks over the teeth of the ratchet wheel. But when the
drum attempts to revolve in the other direction the pawl engages with a
ratchet tooth and prevents the drum from revolving. Some means have
to be provided to lift the pawl out of engagement with the ratchet wheel
in order to permit the load to be lowered.

4.103 Instead of coupling the crank handle directly to the drum of a
winch, a geared arrangement may be used. Toothed gear wheels can be
used to couple any two rotating shafts, and have something in common
with belt coupling in that gear wheels can be used to give a change in
shaft speeds. A gear drive differs from a belt drive in that the shafts

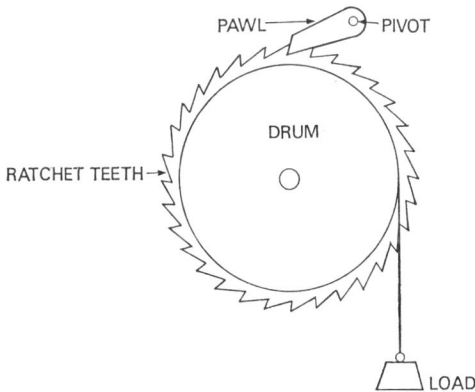

FIGURE 4.20 Ratchet and Pawl.

81

concerned are usually much closer together, so that the gear wheels can mesh properly, also there is no possibility of any slipping of the shafts. The velocity ratio of a pair of gear wheels depends on the number of teeth in each wheel. If the driver wheel makes one revolution the number of teeth on it passing a given point equals the total number on the wheel, but the driven wheel will be forced to revolve so that the same number of its teeth pass a fixed point. It follows that if the driven wheel has twice as many teeth as the driver it will make only one half of a revolution, i.e. the speed of the driven is only half that of the driver. In general the velocity ratio of a pair of gear wheels is given by:

$$\text{V.R.} = \frac{\text{Number of teeth on driven}}{\text{Number of teeth on driver}}$$

This can be expressed as:

$$\frac{\text{Speed of driven}}{\text{Speed of driver}} = \frac{\text{Number of teeth on driver}}{\text{Number of teeth on driven}}$$

If a number of gear wheels are used to form a gear train the resulting overall velocity ratio is the product of the individual velocity ratios, just as was the case for the compound lever discussed in Section 4.73.

Example 4.13

A winch has a drum of diameter 300 mm and a crank of radius 600 mm. The crank turns a shaft to which is fitted a gear wheel having 20 teeth, this meshes with a gear wheel having 600 teeth which is fitted to the drum.
  (a) What is the velocity ratio of the winch?
  (b) What effort is needed to raise a load of mass 200 kg if the efficiency is 60%?
  (c) How many revolutions of the crank handle are required to raise the load by 10 metres?

Solution

  (a) Let the radius of the smaller gear wheel be r. As each wheel must have teeth of similar size it follows that their radii must be proportional to the number of teeth and so the radius of the large gear wheel is $\frac{600}{20} \times r = 30\,r$. The arrangement is similar to a compound lever, the first lever being formed by the crank and first gear wheel and so having V.R. $= \frac{600}{r}$, and the second lever being formed by the second gear wheel and the drum and so having a V.R. $= \frac{30\,r}{150}$.

The overall velocity ratio is given by:

$$V.R. = V.R._1 \times V.R._2$$
$$V.R. = \frac{600}{r} \times \frac{30\ r}{150}$$
$$\underline{V.R. = 120}$$

Alternatively the velocity ratio can be regarded as the original V.R. between crank and drum, modified by the inclusion of the V.R. between the gear wheels. This gives:

V.R. of crank and drum only:

$$V.R._1 = \frac{\text{Radius of crank}}{\text{Radius of drum}}$$
$$V.R._1 = \frac{600}{150}$$
$$V.R._1 = 4$$

(Note that as the drum diameter is 300 mm, its radius is 150 mm.)
V.R. of gear wheels only:

$$V.R_2. = \frac{\text{Number of teeth on driven}}{\text{Number of teeth on driver}}$$
$$V.R._2 = \frac{600}{20}$$
$$V.R._2 = 30$$
$$\text{Overall V.R. of winch} = V.R._1 \times V.R._2$$
$$V.R. = 4 \times 30$$
$$\underline{V.R. = 120}$$

(b) $$M.A. = V.R. \times \text{efficiency}$$
$$M.A. = 120 \times \frac{60}{100}$$
$$M.A. = 72$$
$$\text{Effort} = \frac{\text{Load}}{M.A.}$$
$$\text{Effort} = \frac{200 \times 9\cdot81}{72}$$
$$\underline{\text{Effort} = 27\cdot3\ N}$$

(c) Distance moved by effort = Distance moved by load × V.R.
Distance moved by effort = 10×120 m
Distance moved by effort = 1200 m

As the effort is applied to the crank handle it travels in a circle, completing one circumference of the circle in one revolution. In one revolution the effort moves $2\pi \times 600$ mm, that is 3770 mm or 37·7 m. It follows that the number of revolutions made by the crank is:

$$\text{Revolutions of crank} = \frac{1200}{37 \cdot 7}$$

$$\text{Revolutions of crank} = 31 \cdot 8$$

4.11 Addition of Forces

4.111 When two or more forces act together the resultant force depends not only on the magnitudes of the individual forces but also on their

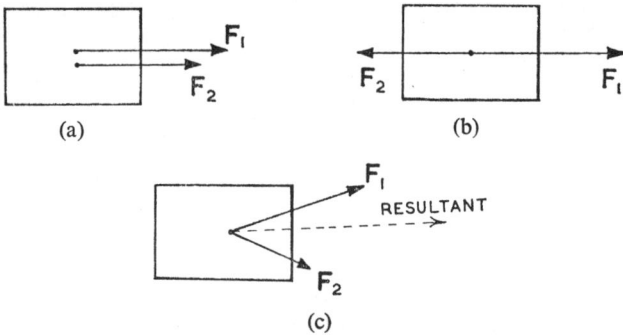

(a)

(b)

(c)

FIGURE 4.21   Forces Acting on a Body.
(a) Forces in line and assisting.
(b) Forces in line and opposing.
(c) Forces not in line.

relative directions. Figure 4.21 shows three situations in which two forces F_1 and F_2 are applied to a body. In the first case, Fig. 4.21(a), the two forces act in the same direction and the resultant is $F_1 + F_2$. In the second case Fig. 4.21(b) the forces are opposing each other, so the resultant is $F_1 - F_2$. In the third case Fig. 4.21(c) the forces are not in line and the resultant will act somewhere between F_1 and F_2 as shown by the dotted line. Also the magnitude of the resultant is obviously greater than either F_1 or F_2, but it is not so great as $F_1 + F_2$ as the two forces are not assisting each other so well as in the first situation.

It is clear that there are two things of importance about a force, its magnitude and its direction of action, and both of these attributes must be taken into account when adding forces. As stated in Section 4.11, quantities like forces, which have both a magnitude and a direction are called *vector* quantities. Any vector quantity can be represented by drawing

84

a line, whose length represents the magnitude of the quantity to a suitable scale, and whose direction shows the direction of the vector quantity.

4.112 It can be seen from the preceding paragraph that the resultant of two forces cannot be found by ordinary arithmetical addition, except in special cases. The resultant can be found, using a scale drawing, by employing the rule of the *parallelogram of forces* which states:

If two forces acting at a point be represented in magnitude and direction by lines forming the two adjacent sides of a parallelogram, then their resultant is represented by the diagonal between them (to the same scale and in the correct direction).

Example 4.14

In order to pull a heavy transformer across a floor two ropes are attached as shown in the diagram (Fig. 4.22) and pulled with the forces

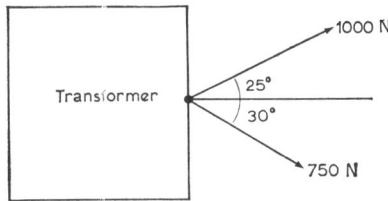

FIGURE 4.22   Example 4.14.

shown. Find the resultant force on the transformer and the direction in which it acts.

Solution

The solution is obtained by drawing the parallelogram of forces to scale as shown in Fig. 4.23. The order in which the drawing should be done is as follows:

Step 1. Choose a suitable scale; this should be chosen so that the diagram is as large as possible without exceeding the size of paper available.

Step 2. Draw the lines representing the forces F_1 and F_2 with their correct length, to scale, and in the correct direction as given in the problem.

Step 3. Complete the parallelogram having the two forces as two of its adjacent sides.

Step 4. Draw the diagonal which lies between the two original forces. This is the resultant, which can now be scaled off to find the required answer. The diagram, shown in Fig. 4.23, gives the solution that the resultant has a magnitude of 1550 N, and acts at an angle of 4° to the reference line as shown.

4.113 In many problems three forces, acting together at a point, are in *equilibrium*. That is, the point is not moving, and any two of the forces are exactly balanced by the third. Problems of this sort can be solved by

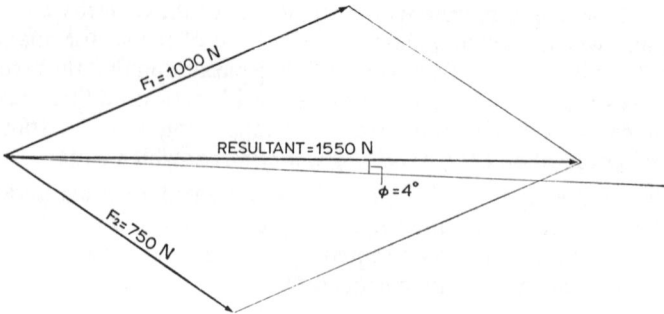

FIGURE 4.23   Solution of Example 4.14 by Scale Drawing Using the "Parallelogram of Forces."

employing the parallelogram of forces and noting that the third force must be equal in magnitude and exactly opposite in direction to the resultant of the first two forces. A more convenient method is to use a modification of the parallelogram rule which is often called the *triangle of forces*. This states that if three forces acting at a point are in equilibrium they may be represented (to scale) by the three sides of a triangle taken in order. When using the triangle of forces principle to solve a problem it is often advantageous to make two diagrams. A *space diagram* which shows the true positions and directions of the forces, and a *force diagram* which is the actual triangle of forces. The sides of the triangle of forces will always be parallel to the directions of the corresponding forces in the space diagram.

Example 4.15

A machine weighing 3600 N is provided with two eye bolts 520 mm. apart. If the machine is lifted by a crane with a chain 720 mm long, as shown in Fig. 4.24, determine graphically the tension in each part of the chain.

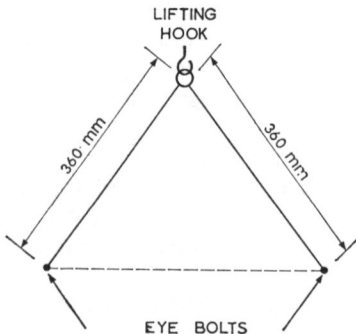

FIGURE 4.24   Example 4.15.

Solution

The problem can be solved by first drawing a space diagram to scale, showing the directions of the forces, and then drawing a triangle of forces. The separate steps required to build up these diagrams are listed below.

Step 1.   Draw a horizontal line and mark the positions of the eye bolts at *A* and *B*.

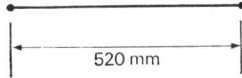

Step 2.   Set a compass to the equivalent of 360 mm radius and draw arcs centred on *A* and *B*.

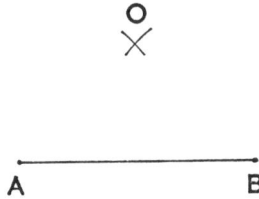

Step 3.   The intersection of the arcs at *O* fixes the position of the lifting hook, the suspension chain *OC* can now be drawn in vertically. The forces acting in the three chains *OA*, *OB* and *OC* all act on the one point *O* and are in equilibrium and so the triangle of forces can be applied as follows:

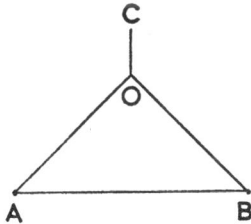

Step 4.   First represent the force in chain *OC* by a line *XY* parallel to *OC* having a length representing 3600 N, as chain *OC* must support the full weight of the machine.

Step 5.   Draw *YZ* parallel to *OA* and *XZ* parallel to *OB*. The intersection at *Z* fixes the lengths of *XZ* and *YZ* so that they can now be scaled off to determine the required answer which is, tension in chain = 2640 N. Figure 4.25 shows the complete construction drawn to scale.

An equally valid alternative to *Step 5* is to draw *YZ* parallel to *OB* and *XZ* parallel to *OA*. The student can easily check that this gives exactly the same result as *Step 5* above. Note that, in this second diagram, the

87

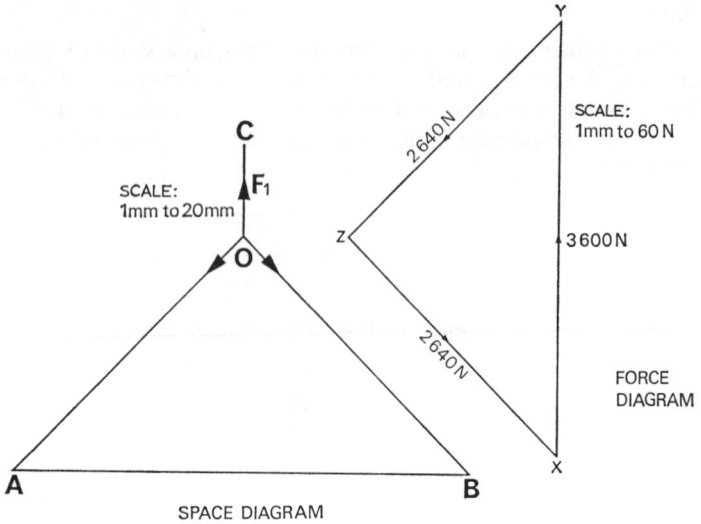

FIGURE 4.25 Solution for Example 4.15.

triangle of forces obtained is the other half of the parallelogram of which
XZ . YZ are the sides and *XY* the diagonal.

Example 4.16

A pivoted arm is linked by a toggle to a fixed pivot point as shown in
Figure 4.26. Determine the force which must be exerted at the toggle
hinge in the direction shown in order to produce a force of 80 N acting
on the pivoted arm.

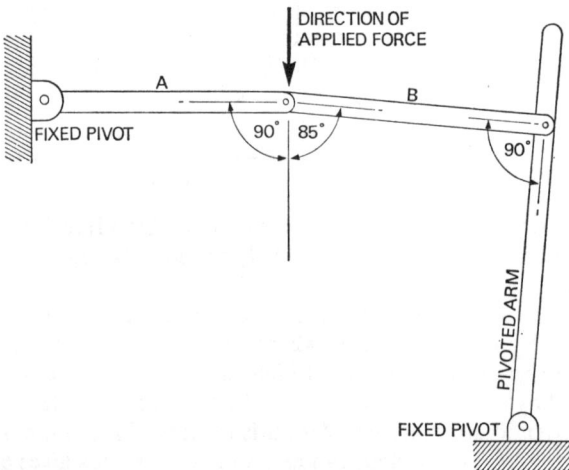

FIGURE 4.26 Example 4.16.

Solution

The force exerted on the pivoted bar is the same as the force acting in the toggle link *B*. The two toggle links, *A* and *B*, and the applied force *F* form a situation in which there are three forces in equilibrium acting at a point, and so the solution may be obtained by a method similar to the preceding example. Figure 4.27 shows the solution drawn to scale, Fig. 4.27(a) being the space diagram and Fig. 4.27(b) being the force diagram.

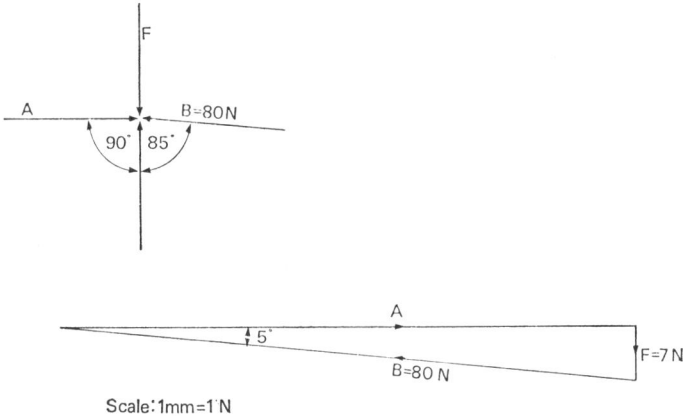

Scale: 1mm=1 N

FIGURE 4.27   Solution for Example 4.16.
(a) Space diagram.
(b) Force diagram.

It is evident that the force diagram is very long and thin in shape. This in turn means that any small error in setting out the angles of the diagram can cause a much larger error in the value of *F* which is the answer required. In such cases it is often worth while to use the methods of elementary trigonometry to solve the force diagram. In this example this is made easy because the force diagram is a right angled triangle, the angle opposite *F* being 5°. It follows that $F = B \times \sin 5°$ (as *B* is the hypotenuse of the triangle) whence

$$F = 80 \times 0\cdot087$$

i.e.
$$F = 6\cdot96 \text{ N}$$

SUMMARY OF FORMULAE

Weight and Mass: Weight $= mg$

Work: Work $=$ force $\times$ distance moved.

Power: Power = rate of doing work

$$\text{Watts} = \frac{2\pi NT}{60}$$

$$\text{Watts} = \frac{\text{newtons} \times \text{metres}}{\text{seconds}} = \frac{\text{joules}}{\text{seconds}}$$

Velocity and acceleration:

$$\text{Speed} = \frac{\text{Distance moved}}{\text{Time taken}}$$

$$\text{Angular velocity} = \frac{\text{Angle turned through}}{\text{Time taken}}$$

$$\text{Acceleration} = \frac{\text{Increase in speed}}{\text{Time taken}}$$

$$\text{Angular acceleration} = \frac{\text{Increase in angular velocity}}{\text{Time taken}}$$

Helical spring: Force = Extension × rate of
                                                              spring

Spiral spring: Torque = Deflection (rad) × rate
                                                   of spring (N m/rad)

Efficiency:

$$\text{Efficiency \%} = \frac{\text{Useful energy output}}{\text{Total energy input}} \times 100$$

$$\text{Efficiency \%} = \frac{\text{Useful power output}}{\text{Total power input}} \times 100$$

$$\text{Efficiency \%} = \frac{\text{M.A.}}{\text{V.R.}} \times 100$$

Velocity Ratio:

$$\text{V.R.} = \frac{\text{Speed of effort}}{\text{Speed of load}}$$

$$\text{V.R.} = \frac{\text{Distance moved by effort}}{\text{Distance moved by load}}$$

$$\text{Compound V.R.} = \text{V.R.}_1 \times \text{V.R.}_2 \times \text{etc}$$

$$\text{V.R. of simple winch} = \frac{\text{Radius of crank}}{\text{Radius of drum}}$$

$$\text{V.R. of gear train} = \frac{\text{Number of teeth on driven}}{\text{Number of teeth on driver}}$$

Mechanical Advantage: $\qquad$ M.A. = $\dfrac{\text{Force acting on load}}{\text{Force applied as effort}}$

Moment and Torque: $\qquad$ Moment = Force $\times$ perpendicular distance from pivot

$\qquad\qquad\qquad\qquad\quad$ Torque = Force $\times$ radius of action.

Belt Drive: $\qquad\qquad\qquad \dfrac{D_1}{D_2} = \dfrac{N_2}{N_1} = \dfrac{T_1}{T_2}$

Gear Drive: $\qquad \dfrac{\text{Speed of driven}}{\text{Speed of driver}} = \dfrac{\text{Number of teeth on driver}}{\text{Number of teeth on driven}}$

EXERCISE 4

1. (a) Define (i) power (ii) efficiency
   (b) A hoist is designed to raise a load of 1200 kg through a height of 50 m in 20 seconds. If the efficiency of the hoist mechanism is 70% what is the power output of the motor in kW?
   (c) If the motor efficiency is 82% what current does it take from a 500 V d.c. supply?

   (1 kg weighs 9·81 N)

2. Describe, in general terms, an engineering application where:
   (a) heat energy is converted into mechanical energy;
   (b) mechanical energy is converted into heat energy.
   An electric motor drives a pump which on full load raises water through a height of 10 m at 90 litres per minute. Determine the power of the pump and the power output of the motor if the pump efficiency is 90%. (1 litre of water weighs 9·81 N.)

3. Give and explain the relationship between work and power. An electric motor runs at 1000 rev/min and produces a torque of 99 N m. Calculate the power of the motor and the electrical power input, if the motor efficiency is 80%.

4. A 240 V d.c. motor driving a pump takes a current of 15 A when raising 400 litre of water to a height of 15 m in 30 seconds. Calculate the overall efficiency of motor and pump.
   (1 litre of water weighs 9·81 N).

5. A hoist lowers a load of 4 t slowly through a height of 30 m under the control of brakes. Calculate the energy absorbed by the brakes (1 t = 1000 kg and g = 9·81 m/s².)

6. During a brake test of a d.c. motor the torque developed was equivalent to a force of 25 N acting at a radius of 0.5 m. The motor speed was 1200 rev/min and the motor took a current of 9 A from a 220 V d.c. supply. Determine the efficiency of the motor for the conditions of the test.

7. The overall efficiency of a portable electric pumping set is 60 per cent and it works from a 240 V supply. Calculate the current taken by the set when it is pumping 450 litres of water per minute from a depth of 5 m. (1 litre of water weighs 9·81 N.)

8. Explain the meaning of torque. An electric crane lifts a load of 4 t at a speed of 550 mm/s. It is operated by a 400 V motor, the current taken being 80 A. Calculate the efficiency of the motor. (1 t = 1000 kg and g = 9·81 m/s².)

9. Explain the term *moment of force* about a point. A piece of steel wire is to be cut by means of a pair of wire cutters. The wire is gripped 5 mm away from the fulcrum, whilst the pressure is applied to the handles 80 mm from the fulcrum. What force must be applied to the handles to cut the wire if it requires a direct cutting force of 160 N.

10. A set of pulley blocks having 3 pulleys in each sheave is used to lift a load weighing 100 N. The pull required on the free end of the rope is 25 N. Make a diagrammatic sketch of the tackle and determine its velocity ratio, mechanical advantage and efficiency.

11. A machine is driven by vee-belt from an electric motor which runs at 1440 rev/min. The motor is fitted with 60 mm diameter pulley and the machine with 240 mm diameter pulley. The torque required to drive the machine is 8 N m. Assuming no belt slip, and an overall efficiency of 80%, calculate
    (i) the speed of the machine
    (ii) the power consumed by the motor.

12. Explain the principle of the *parallelogram of forces*. Two forces of 3 N and 4 N act at right angles to each other. Find, by drawing a parallelogram, the resultant of these two forces. Measure the angle between the resultant and the 4 N force.

13. Define the term *moment of force* and illustrate your answer with the aid of a diagram.
    Three forces $OA = 8$ N, $OB = 12$ N, and $OC = 16$ N, all act outwards from point O. The angles between the direction of the forces are $AOB = 80°$, $BOC = 120°$, and $COA = 160°$. By drawing to a suitable scale find the *resultant* of these forces both in magnitude and direction.

5 Temperature and Heat

5.1 Temperature

5.11 The *temperature* of an object is a measure of how hot it is. Temperatures are measured using *thermometers*, a common form of thermometer being the mercury in glass type illustrated in Fig. 5.1. The instrument consists of a stem in the form of long, fine-bore glass tube or

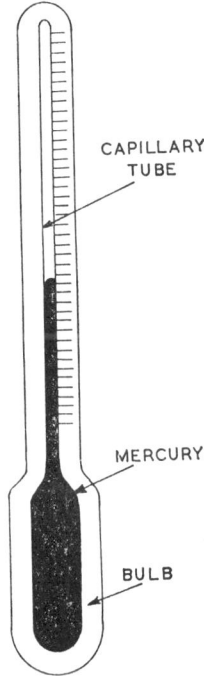

FIGURE 5.1   Mercury in Glass Thermometer.

capillary tube, the lower end of which is formed into a bulb. The bulb and part of the stem are filled with mercury. When the thermometer is heated the mercury expands, and the level of mercury in the stem increases so indicating the temperature.

5.12 The SI unit of temperature measurement is the kelvin (abbreviation K). The Kelvin temperature scale has its zero point at "absolute zero", that is the lowest temperature which can theoretically be obtained. As this temperature scale is not very convenient for everyday use the non-SI unit, the degree Celsius (abbreviation °C) is widely used for many purposes. The Kelvin and Celsius scales have equally spaced temperature intervals, they differ only in the position of the zero mark as shown in Figure 5.2. Thus to convert a value in °C into the corresponding value in K it is only necessary to add the constant 273

$$K = °C + 273$$

Similarly it follows that:

$$°C = K - 273$$

Thermometers are often calibrated by reference to two *fixed points*, i.e. two temperatures which can easily be reproduced accurately. The lower fixed point, often called the freezing point, is the temperature at which ice melts, while the upper fixed point, often called the boiling point, is the temperature at which water boils under normal atmospheric pressure. Freezing point is equal to 273 K or 0°C and boiling point is equal to 373 K or 100°C as indicated in Figure 5.2. In many cases the result of a calculation will give the rise or fall of temperature in a certain situation. Figure 5.2 shows that between freezing and boiling points the Kelvin scale has a temperature difference of 373 K − 273 K = 100 K, while the Celsius scale has a temperature difference of 100°C − 0°C = 100°C. It can be seen that: Temperature rise (or fall) in K = Temperature rise (or fall) in °C

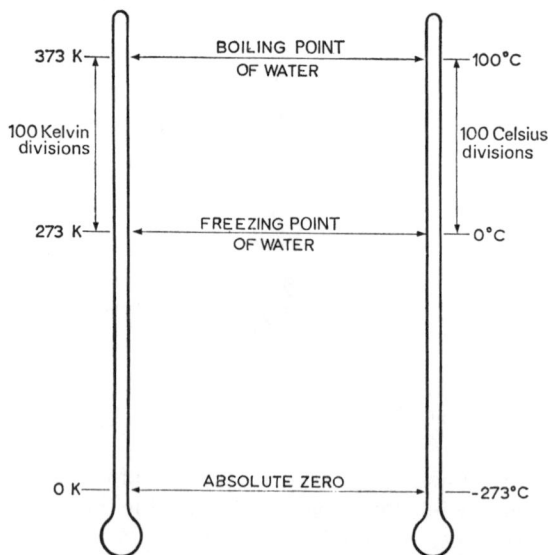

FIGURE 5.2 Temperature Scales.

5.2 Heat

5.21   Heat is a form of energy. When heat is added to a substance one of two things may happen:

(a) the temperature of the substance may increase,
or   (b) the state of the substance may change.

For example, suppose some ice at a temperature below 0°C is heated. At first the temperature rises until it reaches 0°C; the ice then starts to melt and adding further heat causes no further rise in temperature until all the ice has melted. The heat added during this period is used to change the state of the substance from ice into water. After the ice has melted adding further heat causes the temperature of the water to rise until a temperature of 100°C is reached. The addition of further heat now converts the water to steam.

5.22   The SI unit of energy is the joule and, as heat is a form of energy, it follows that quantities of heat are measured in joules. The amount of heat in joules needed to raise the temperature of a 1 kg mass of substance by 1 K (or 1°C) is called the *specific heat capacity* (symbol c) of the substance. For example water has a specific heat capacity of 4180 joules per kilogramme kelvin, meaning that it requires 4180 joules to raise the temperature of 1 kg of water by 1 K (or 1°C). The amount of heat needed to cause a given temperature rise is given by:

Heat energy (in joules) = Mass (in kg) × specific heat capacity (in J/kg K) × temperature rise (in K or °C)

$$W = m \times c \times (\theta_2 - \theta_1)$$

Example 5.1

How much heat is required to raise the temperature of 20 litres of water from 15°C to 55°C? One litre of water has a mass of 1 kg.

Solution

$$W = m \times c \times (\theta_2 - \theta_1)$$
$$W = 20 \times 4180 \times (55 - 15)$$
$$W = 3\ 344\ 000 \text{ J} \quad \text{or} \quad 3.344 \text{ MJ}$$

5.23   Heat quantities are sometimes measured in *calories* (Symbol Q abbreviation cal). The calorie is a non-SI unit which is defined as follows:
One calorie is the amount of heat required to raise the temperature of a mass of one gramme of water by 1 °C. It follows that as the specific heat capacity of water is 4180 J/kg K one calorie is equivalent to 4.18 joule. Thus to convert calories to joules it is only necessary to multiply by 4.18 (a widely used approximate value is 4·2).

5.24 The table below gives the specific heat capacities of some commonly used materials.

Specific Heat Capacities	
Material	SI Value J/kg
Water	4180
Copper	390
Brass	380
Steel	450
Aluminium	880
Paraffin Oil	2180
Transformer Oil	2130

The values of specific heat capacity given above vary slightly with temperature and, in the case of alloys such as brass or steel, with composition. It is convenient, and sufficiently accurate for most purposes to take the specific capacities of copper, brass and steel as being equal to 400 J/kg K.

5.3 Electrical Heating

5.31 Heat is often produced by converting electrical energy, for example by passing an electric current through a resistor. As with any other machine the heat conversion process is seldom 100% efficient. For example, in the case of an electric kettle, although all the electrical energy input is converted to heat, some of this heat is lost from the kettle and warms the surroundings rather than the water in the kettle; also the metal of the kettle itself is heated as well as the water in it. The efficiency of a heating process can be increased by reducing the heat losses. There are two principal methods of doing this, namely by using heat insulation or lagging to retain the heat, and by reducing the time taken to produce the heat, so reducing the time during which losses can take place. A good example of the latter method is afforded by the modern high speed electric kettle in which the heating element is rated as highly as possible while still being able to operate from a standard 13 A socket outlet. The high rating ensures quick boiling of the kettle and this in turn helps to maintain a high efficiency.

Example 5.2

(a) What is the difference between *heat* and *temperature*?

(b) Calculate the time required for an electric water heater to raise the temperature of 10 litres of water from 15°C to 65°C, if the heating element takes a current of 12 amperes from a 240 V supply and the efficiency of the heating process is 85%. Assume the specific heat capacity of water to be 4180 J/kg K.

Solution

(a) Heat is a form of energy which, when added to a substance, causes it to become hotter, whereas temperature is a measure of the degree of hotness of the substance. The temperature indicates how hot the substance is, but not how much heat it contains, as the quantity of heat depends not only on how hot the substance is but also on how much there is of it.

(b) The amount of heat energy required is given by:

$$W = mc(\theta_2 - \theta_1)$$
$$W = 10 \times 4180 \times (65 - 15)$$
$$W = 2\,090\,000 \text{ J}$$

Allowing for the efficiency:

$$\text{Electrical energy required} = 2\,090\,000 \times \frac{100}{85}$$
$$W = 2\,459\,000 \text{ J}$$
$$\text{Electrical power input,} \quad P = E \times I$$
$$P = 240 \times 12$$
$$P = 2880 \text{ W}$$
$$\text{Time taken,} \quad t = \frac{W}{P}$$
$$t = \frac{2\,459\,000}{2880}$$
$$t = 854 \text{ s}$$
$$t = 14 \text{ min } 14 \text{ s}$$

5.4 Transmission of Heat

5.41 There are three methods by which heat can flow from one place to another, namely conduction, convection and radiation. Heat is said to

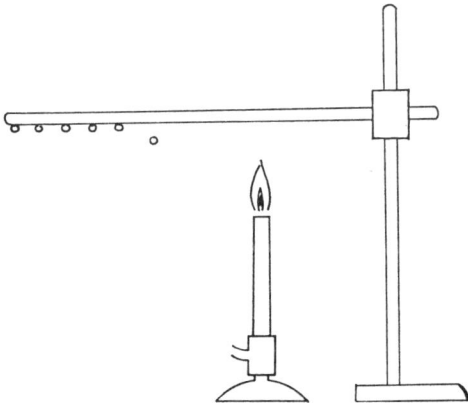

FIGURE 5.3   Experiment to Show Heat Conduction.

be transmitted by *conduction* when it actually flows through a material. Figure 5.3 shows a metal rod with several small ball bearings initially stuck to the rod by means of a thin coating of paraffin wax. As the end of the rod is heated, the heat is conducted along the rod and this is shown by the ball bearings dropping off one after the other as the wax melts under the heat transmitted through the rod. Note that the ball bearings nearer the source of heat drop off first. It is clear from this that the transmission of heat through the rod by means of conduction takes time. The rate of transmission of heat in the substance could be assessed by measuring the time taken for two adjacent ball bearings to drop off.

5.42 Heat is said to be transmitted by *convection* when the heat is carried along by the actual movement of heated particles. For example, in a convector heater cold air is drawn in and is heated by the heating element. As the temperature of the air is increased it expands, becomes lighter and so tends to rise. The result is a continuous current of warm air out of the top of the heater and a corresponding flow of colder air into the bottom of the heater as illustrated in Fig. 5.4(a). Currents of this type which can be set up in any fluid or gas, are called *convection currents*. Figure 5.4(b) shows an example of a convection current in water. In this case an immersion heater, fitted at the bottom of a tank warms the water in its vicinity. This water then rises and creates a convection current which distributes the heat around the tank. As water is a poor conductor of heat most of the heat transmission throughout the tank is by the convection

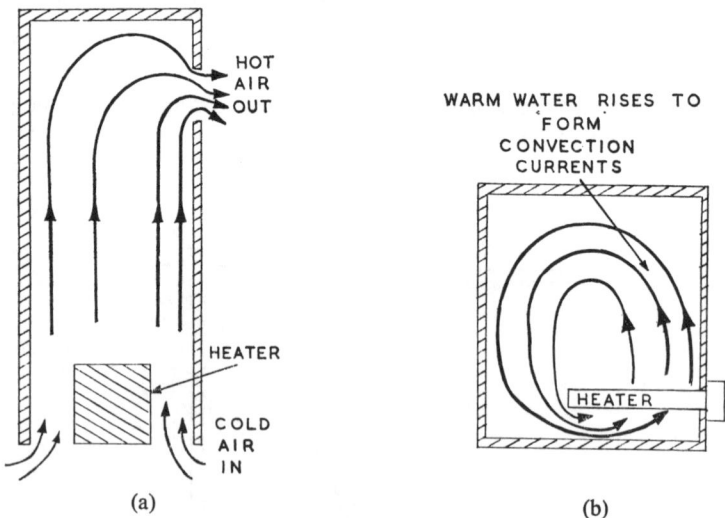

FIGURE 5.4   Convection Currents.
(a) Convector heater.
(b) Immersion heater.

effect, the water below the immersion heater receiving only a little of the heat. It is for this reason that immersion heaters are usually fitted fairly low in the tanks which they heat.

5.43 Heat can also be transmitted in the form of infra-red rays which are similar in many respects to light rays. Transmission of heat in this manner is known as *radiation*. The rays of heat have the property that they do not warm the substance through which they pass, they warm only the object which absorbs them. The heat rays can be reflected and focussed just as can rays of light and they can pass through empty space, i.e. they can pass through a vacuum, indeed this is how heat from the sun reaches the earth. It is found that objects having a dull surface can emit and absorb radiant heat readily, whereas objects with a shiny surface neither emit nor absorb radiant heat easily.

5.44 In order to keep the efficiency of equipment such as water heaters as high as possible, the heat losses must be reduced to a minimum. Heat loss by conduction can be reduced by lagging, that is by enclosing the equipment in a jacket of poorly conducting material, such as glass fibre, slag wool or granulated cork. Radiation losses can be reduced by using a shiny outer surface.

Example 5.3

A tank containing 60 litres of fuel oil is heated by means of a 750 W immersion heater. Assuming that the overall efficiency of the operation is 80%, calculate the time required to raise the temperature of the oil from 10°C to 35°C. Explain how the efficiency of the operation could be improved.

One litre of fuel oil has a mass of 0·85 kg.

The specific heat capacity of the oil is 1900 J/kg K.

Solution

Heat energy required, $W = mc(\theta_2 - \theta_1)$
$$W = 60 \times 0.85 \times 1900 \times (35 - 10)$$
$$W = 2\,422\,500 \text{ J}$$

Allowing for efficiency:

Electrical energy required, $W = 2\,422\,500 \times \dfrac{100}{80}$

$$W = 3\,028\,000 \text{ J} \quad \text{or} \quad 3\cdot028 \text{ MJ}$$

Time taken, $t = \dfrac{W}{P}$

$$t = \dfrac{3\,028\,000}{750}$$

$$t = 4028 \text{ s}$$

$$t = 1 \text{ h } 7 \text{ min (to the nearest minute)}$$

The efficiency of the operation can be improved by reducing the heat losses from the oil tank. This can be done by lagging the tank, that is by covering it with a heat insulating material such as a glass fibre jacket.

5.5. Expansion

5.51 Many materials expand when heated. This effect can be utilised to measure temperature as in the mercury thermometer already described, in which the amount of expansion of the mercury is used to indicate the temperature. The expansion effect can be put to good use, for example, when fitting certain types of bearing into a casing. The casing is bored out very slightly smaller than the outer diameter of the bearing, and then

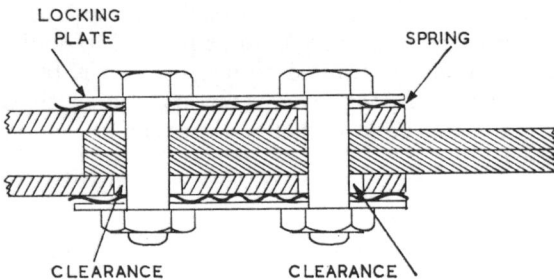

FIGURE 5.5 Expansion Joint for Bus-Bar Trunking.

heated so that it expands. This enables the bearing to be placed in position in the hole, so that when the casing cools and shrinks back to its original dimensions the bearing is tightly gripped. Expansion can also cause difficulties in certain cases, for example bridges are often supported on rollers to allow for expansion, and gaps have to be left between sections of railway lines. Figure 5.5 shows an expansion joint for use in a long run of bus-bar trunking. Here the use of such a joint helps to prevent the copper bus-bars buckling when their temperature increases. When installing overhead lines in hot weather care has to be taken to allow sufficient sag in the line otherwise, when colder weather arrives, the line in contracting will become so taut that it may snap.

5.52 Thermostats are temperature operated switches. They are often used to provide automatic temperature control of the heating apparatus to which they are fitted, by switching off the heating device when the temperature rises above the required level and switching on again when the temperature falls below the required level. Figure 5.6 shows a type of thermostat which relies on the different rates of expansion of the two metals. The outer tube is made of brass which expands by a small amount when heated, the inner rod is made of invar steel which does not expand when heated. The brass tube and rod are joined at the end remote from

100

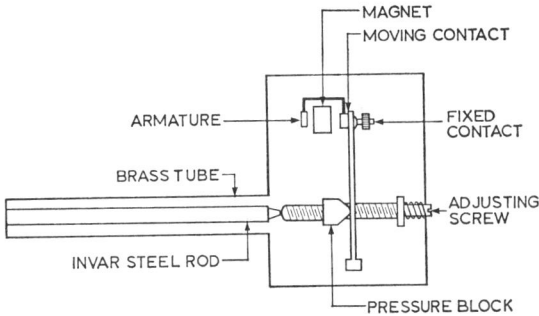

FIGURE 5.6   Water Heater Thermostat.

the switch, so that as the temperature rises the expansion of the brass tube causes a reduction of the pressure applied to the contact spring and eventually the contacts separate. The small permanent magnet provides a snap action to the switch. This type of thermostat is often used in water heaters. Figure 5.7(a) shows an air thermostat which uses a bimetallic strip as the temperature sensing element. The bimetallic strip Fig. 5.7(b) consists of two strips of different metals, often brass and steel, joined face

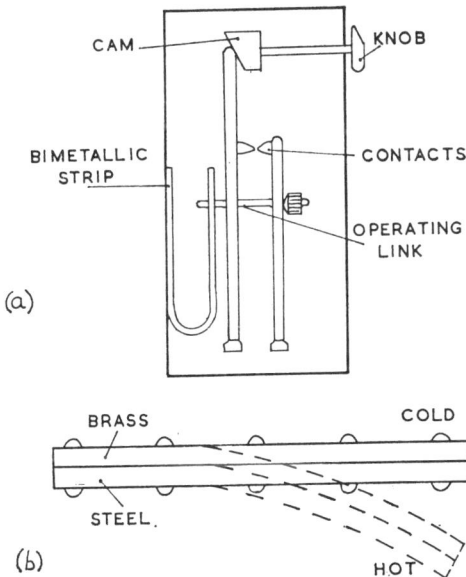

FIGURE 5.7   Air Thermostat.
(a) Air thermostat.
(b) Bimetallic strip.

101

to face either by rivetting or welding. When the temperature rises the brass expands more than the steel and this causes the strip to bend. The oven thermostat shown in Fig. 5.8 relies on the evaporation of a volatile

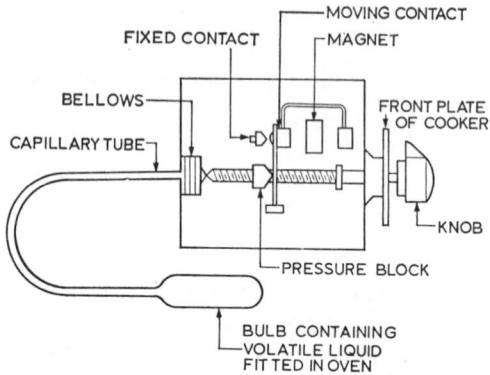

FIGURE 5.8   Oven Thermostat.

liquid. As the temperature rises, the liquid in the bulb tends to evaporate so raising the pressure inside the bulb. This pressure is transmitted via the capillary tube to the bellows in the switch unit, and so operates the switch. This type of thermostat has the advantage that the temperature sensing bulb can be mounted at some distance from the switch when required.

SUMMARY OF FORMULAE

Conversion of temperature scales: $K = °C + 273$

$$°C = K - 273$$

Heat Energy: $W = mc(\theta_2 - \theta_1)$

EXERCISE 5

1. An electric water heater raises the temperature of 12 litres of water from 12°C to 55°C in 30 minutes. If the supply voltage is 240 V calculate the current taken from the supply, assuming that there are no heat losses.

    (1 litre of water has a mass of 1 kg. The specific heat capacity of water may be taken as 4·2 kJ/kg K.)

2. (a) State the three methods by which heat can be transmitted.
   (b) Draw a good diagram and describe fully how an electric immersion heater heats the water in a hot water cylinder. Why should such a cylinder be lagged and what type of material should be used to do this?   U.E.I.

3. Describe three ways by which heat may be transferred. Give an example of the use of each of these principles in electrical apparatus. An electric immersion heater is required to heat 80 litres of water from 15°C to 65°C in 80 minutes Calculate:
   (a) the heat energy required in joules
   (b) the rating of the heater if there are no losses
   (c) the cost at 0·5p per kW h.
   (1 litre of water has a mass of 1 kg. Specific heat capacity of water = 4200 J/kg K.)

4. A 3 kW immersion heater is installed in a 100 litre water tank. If the initial temperature of the water is 15°C calculate the time taken to heat the water to 70°C assuming that the heating process is 80% efficient.
   (1 litre of water has a mass of 1 kg. Specific heat capacity of water = 4·2 kJ/kg K.)

5. (a) A water heater holding 50 litres of water is heated by means of a 4 kW immersion heater. The time taken to raise the temperature of the water from 15°C to 68°C is 54 minutes. Calculate the efficiency of the operation.
   (1 litre of water has a mass of 1 kg. Specific heat capacity of water = 4·2 kJ/kg K.)
   (b) Make a simple sketch of a water heater tank in the shape of a cube, and show where you would fix an immersion heater. Explain the reasons governing your choice of the position of the immersion heater.

6. An electric water heater contains 3 kg of water and is to be used on a 140 V supply. It is required to raise the temperature of the water from 20°C to boiling point in 10 min. Assuming an efficiency of 85%, calculate:
   (a) the cost of boiling this water if electrical energy is 0.7p per kW h;
   (b) the current.

7. (a) What is meant by the "specific heat capacity" of a substance?
   (b) Two kilogrammes of lead at a temperature of 15°C are to be melted. If the melting point of lead is 325°C and its specific heat capacity is 1350 J/kg K, calculate the amount of heat which must be supplied.

8. (a) Define *specific heat capacity*.
   (b) A quantity of oil weighing 20 kg and of specific heat capacity 3 kJ/kg K is to be heated electrically from 15°C to 50°C. Calculate the number of joules required.

9. In what three ways can heat be transferred? Give a practical example of each.
   An electric soldering iron is heated from 240 V mains and takes a current of 0·5 A. The mass of the copper bit is 120 grammes and 60% of the heat generated is wasted in heating the other metal parts of the iron and in radiation. If the room temperature is 15°C, find the time taken for the iron to reach a temperature of 315°C. (Take the specific heat capacity of copper as 420 J/kg K.)

10. Explain how heat may be transmitted from a heated object. An insulated tank containing 300 litres of oil is heated by means of a 3 kW immersion heater. The oil temperature rises from 20°C to 60°C in a time of 2½ hours. Find the efficiency of the operation. (1 litre of oil has a mass of 0·9 kg. The specific heat capacity of the oil is 2 kJ/kg K.)

11. Describe and explain the operation of the thermostat as used for controlling the temperature in an electrically-heated water tank, A 1500 W immersion heater is used to heat 45 litres of oil. If the efficiency of the operation is 75 per cent, find the time needed to raise the temperature of the oil from 20°C to 70°C (1 litre of oil has a mass of 0·9 kg, the specific heat capacity of the oil is 2050 J/kg K.)

6 Magnetic and Electric Fields

6.1. Magnetic Fields

6.11 One of the most obvious properties of a magnet is its ability to attract to itself other pieces of iron or steel. Materials which can be attracted by a magnet are called magnetic materials while those materials which are not attracted by a magnet are called non-magnetic materials. The most common magnetic materials are iron and steel while such materials as copper, brass, paper and wood are non-magnetic. Simple experiments show that the magnetic effects of a magnet appear to emanate from *poles* which, in the case of a bar magnet, are located near to each end. A freely suspended bar magnet tends to point approximately North-South. The pole which points towards the North is generally called the North-seeking pole, while the pole which points towards the South is called the South-seeking pole. This effect is utilised in the magnetic compass. When two magnets are brought together it is found that like poles repel each other while unlike poles attract.

6.12 The influence of a magnet extends through space, and passes through non-magnetic materials. For example, if a magnet is placed on one side of a sheet of paper it can attract a steel pin placed on the other side of the paper. The region of space throughout which its influence extends is called the magnetic field of the magnet. Magnetic fields may be mapped by drawing lines which indicate the direction in which an isolated North pole would move under the influence of the field if it were free to do so. As isolated North poles cannot be obtained in practice, magnetic fields can be plotted by noting that a short compass needle will set itself in the direction of the field lines at the point where it is placed. A simple method of obtaining a map of the field set up by a magnet is to place the magnet under a sheet of paper, or glass, and to sprinkle iron filings on top of the paper (or glass). After gentle tapping the filings will be seen to have arranged themselves into a map of the field. By studying the fields produced by magnets under various conditions it is found that:

    (a) The magnetic field lines never cross.

    (b) The lines can be continued through the body of the magnet producing the field and they then form complete loops.

    (c) The part of the field line outside the magnet runs from a North to a South pole.

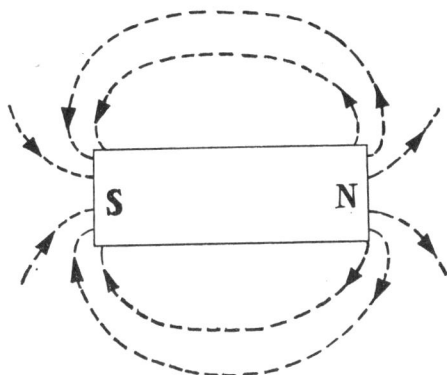

FIGURE 6.1   Magnetic Field of Bar Magnet.

6.13   The field set up by a magnet is permeated with *magnetic flux* (symbol Φ), and the stronger the magnet the more flux it produces. Magnetic flux is measured in **webers** (abbreviation Wb). The strength of the magnetic field at any point is the concentration of the magnetic flux at that point; this is called the *flux density* (symbol B) at the point. Flux density is measured in webers per square metre (abbreviation Wb/m²); an alternative name for this unit is the **tesla** (abbreviation T).

$$B \text{ (magnetic flux density)} = \frac{\Phi \text{ (magnetic flux)}}{A \text{ (area through which flux passes)}}$$

Example 6.1

The poles of an electric motor each have a cross-sectional area of 0·6 m². Calculate the magnetic flux density in the poles when the total flux per pole is 0·66 Wb.

Solution

$$B = \frac{\Phi}{A}$$

$$B = \frac{0·66}{0·6}$$

$$B = 1·1 \text{ Wb/m}^2 \text{ or } 1·1 \text{ T}$$

6.2   Electromagnets

6.21   Whenever an electric current flows through a conductor a magnetic field is set up. The shapes of the magnetic fields produced by various arrangements of conductor are shown in Fig. 6.2.

105

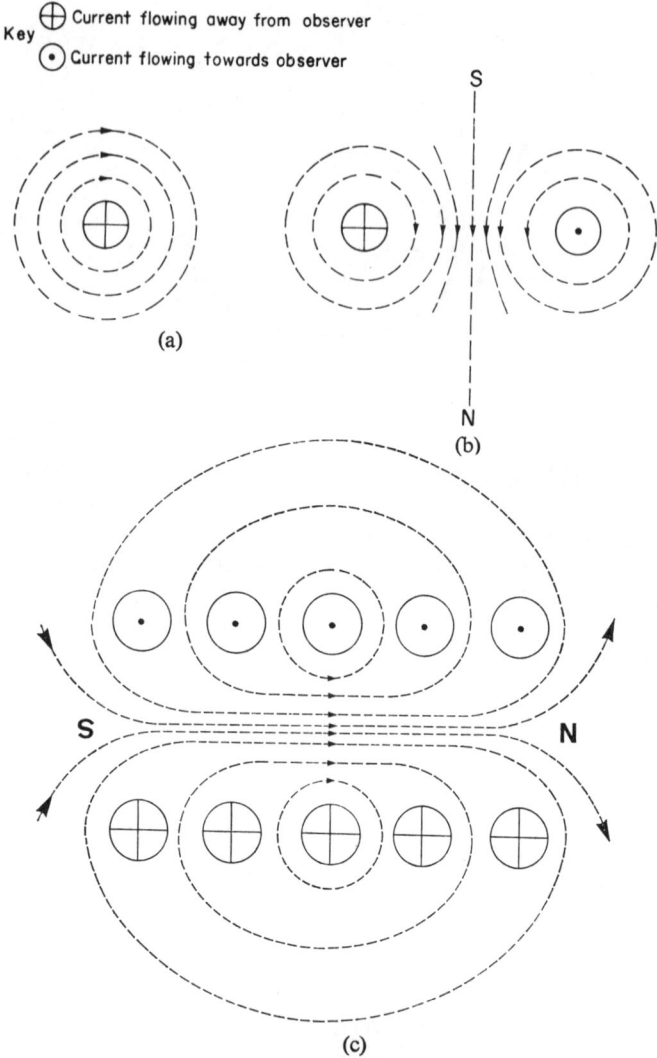

Key ⊕ Current flowing away from observer
⊙ Current flowing towards observer

FIGURE 6.2 Magnetic Fields due to Electric Currents.
(a) Straight conductor.
(b) Flat coil.
(c) Long coil or solenoid.

6.22 The direction of the magnetic field lines produced by a current in a straight conductor may be found by noting that, when looking along the conductor in the direction in which the current flows, the field lines form concentric circles with a clockwise rotation. This is known as Maxwell's corkscrew rule as, if a corkscrew is turned with the same

106

rotation as the magnetic lines, it will screw itself along in the direction of the current. The field set up by a long coil, or solenoid, is similar to that of a bar magnet, and its polarity may be found by considering the direction of the flux set up by the current in each part of the conductor, using the corkscrew rule. A good alternative is to use the right hand grip rule which states that, if the coil is gripped in the right hand with the fingers pointing in the direction of the current, then the thumb points to the North pole of the coil. Another useful rule is the "end" rule illustrated by Fig. 6.3. If the end of the coil is viewed the direction of the current forms the arms of either an S or an N, so giving the polarity of that end of the coil.

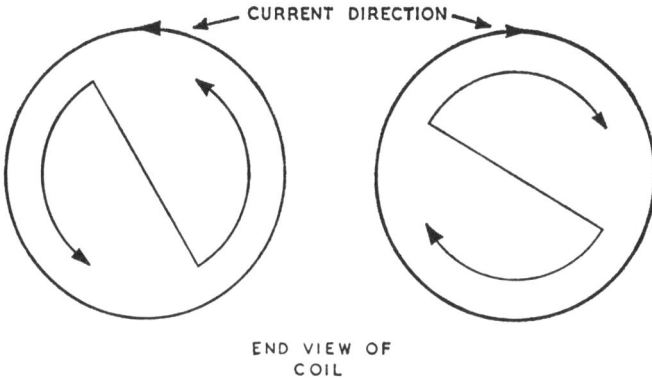

FIGURE 6.3 "End" Rule for Determining Magnetic Polarity.

6.23 The strength of the magnetic field produced by a coil can be greatly increased by inserting a core of magnetic material in it. This is because the magnetic material itself becomes magnetised and so helps to increase the field strength. It is found that some materials are easily magnetised but lose their magnetism easily. Such materials, which include soft iron, silicon steel and stalloy are called *magnetically soft*. They are used for the cores of electro magnets, which are required to lose their magnetism as soon as the current is switched off. Other materials exist which, although often more difficult to magnetise in the first place, retain most of their magnetism after the current is switched off. These materials, which include cobalt steel, alnico and ticonal, are called *magnetically hard*, and they are used to make permanent magnets.

6.3 Magnetic Circuits

6.31 A magnetic circuit is an arrangement for guiding magnetic flux around a definite path, so that the flux produced in the arrangement is concentrated at the point where it is to be used, only a very small fraction

107

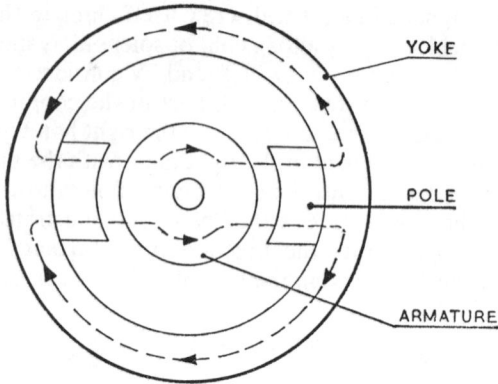

FIGURE 6.4   Magnetic Circuit of 2-Pole Motor.

of the total flux leaking out of the useful path. Figure 6.4 shows the simplified magnetic circuit of a two pole electric motor.

6.32   The force producing magnetic flux in a magnetic circuit is called the *magnetomotive force* (abbreviation m.m.f. symbol F). The m.m.f. can be provided by a permanent magnet but, more often, it is provided by a coil carrying an electric current. The m.m.f. produced by a coil depends both on the current flowing and on the number of turns: the SI unit of m.m.f. is the **ampere** (abbreviation A), but the non-standard unit **ampere turn** (abbreviation A t) is widely used whenever it is desirable to distinguish clearly between the units of m.m.f. and current.

$$F \text{ (m.m.f.)} = I \text{ (amperes)} \times N \text{ (number of turns)}$$

6.33   The influence of the m.m.f. is spread over the whole length of the magnetic circuit. It is however often necessary to determine the effect at some particular point in the circuit. For this purpose the *magnetizing force* (or *magnetic field strength*) (symbol H) at the point concerned is required. If a magnetic circuit is completely uniform, that is composed of the same material and of constant cross-sectional area throughout, then the magnetizing force throughout the whole circuit is given by

$$H \text{ (magnetizing force)} = \frac{F \text{ (m.m.f.)}}{l \text{ (length of circuit in metres)}}$$

and as $F = IN$ it follows that:

$$H = \frac{IN}{l}$$

and that the unit in which H is measured is the ampere turn per metre (abbreviation A t/m), the strict SI unit being the ampere per metre (abbreviation A/m).

Example 6.2

A coil of 200 turns is wound on a uniform magnetic circuit of length 0·25 metres. Determine the m.m.f. and the magnetizing force when a current of 2 amperes is flowing through the coil.

Solution

$$F = IN$$

$$F = 2 \times 200$$

$$\underline{F = 400 \text{ A t}}$$

$$H = \frac{IN}{l}$$

$$H = \frac{2 \times 200}{0·25}$$

$$\underline{H = 1600 \text{ A t/m}}$$

6.4 Magnetization Curves

6.41 The flux density set up in any material depends both on the magnetizing force acting and on the type of material. The relation between flux density and magnetizing force can be shown by means of a graph called a *magnetization curve* or *B/H* curve. Figure 6.5 shows the magnetization curves for some commonly used magnetic materials.

6.42 It can be seen from Fig. 6.5 that the *B/H* curve rises steeply at first and then tends to flatten out. This flattening of the curve is due to *magnetic saturation*. All magnetic material is easily magnetized up to a certain level, but then the material becomes saturated or full of magnetism and it becomes increasingly more difficult to increase the flux density any further. This is indicated by the flattening of the *B/H* curves to the right.

6.43 If the magnetizing force applied to a piece of magnetic material is increased up to the saturation level and then decreased, the values of flux density obtained with a falling magnetizing force are greater than those obtained with a rising value. This effect is known as magnetic *hysteresis*. Figure 6.6 shows the effect of varying the magnetizing force from a positive to a negative value and back again. The curve so obtained is called a hysteresis loop.

6.44 It is the hysteresis effect which causes a magnetic material to retain some of its magnetism after the magnetizing force has been removed. This *residual magnetism* is very important for the satisfactory operation of self-excited dynamos, as explained in Chapter 7. "Hard" magnetic materials

109

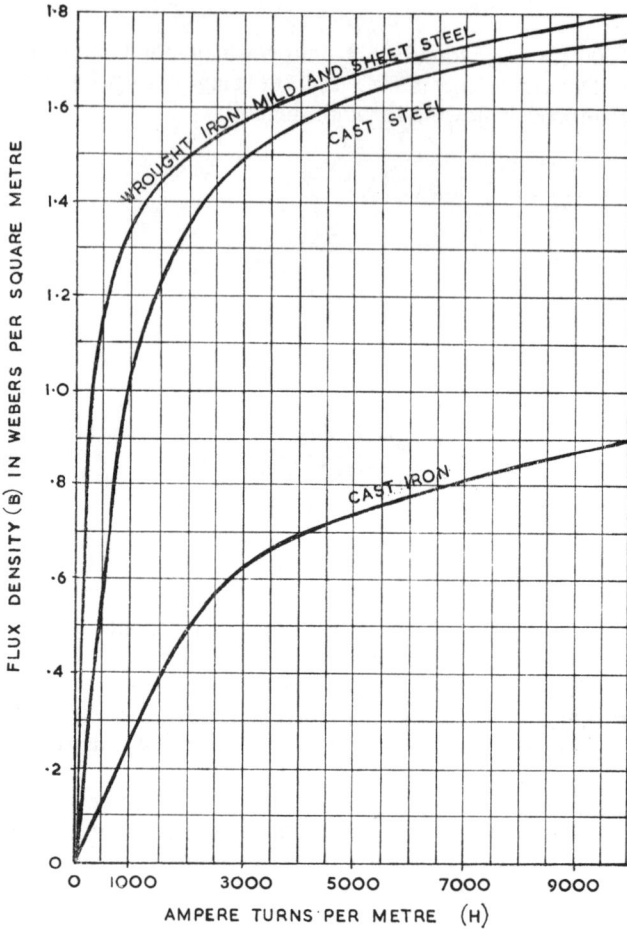

FIGURE 6.5   Magnetization Curves.

have a wide hysteresis loop and retain a high level of magnetism as the hysteresis loop cuts the *B*-axis at a high value. "Soft" magnetic materials are more easily magnetized but do not retain so much of their magnetism as the hysteresis loop cuts the *B*-axis at a low value. The hysteresis effect causes an energy loss whenever the magnetic flux changes, the amount of loss being proportional to the area of the hysteresis loop. It follows that for applications such as transformer cores, where the flux is alternating, a material with a narrow hysteresis loop is required. The table on page 111 lists the main properties of some commonly used magnetic materials.

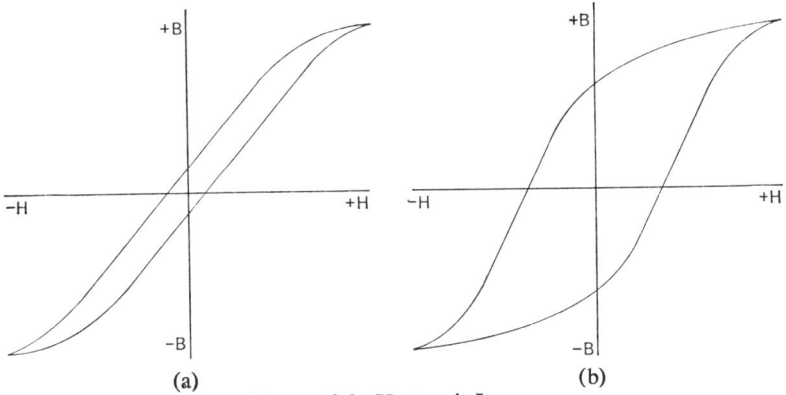

FIGURE 6.6 Hysteresis Loops.
(a) Magnetically soft material.
(b) Magnetically hard material.

Magnetic Materials		
Material	*Principal Magnetic Property*	*Application* .
Cobalt steel	Retains magnetism strongly	Permanent magnets
Alnico	Retains magnetism better than cobalt steel	Powerful permanent magnets
Soft iron	Easily magnetized and demagnetized	Relay cores, electric bell magnets
Silicon iron	High permeability, low hysteresis loss	Transformer cores, motor armatures, etc.
Mumetal	Very high permeability	Special transformers

Example 6.3

A magnetic circuit, in the form of a mild steel ring having a mean diameter of 200 mm and a cross-sectional area of 500 mm², is wound with a coil of 500 turns. Calculate the value of current which must flow through the coil in order to produce a total flux of 0·5 mWb.

Solution

It is necessary first to calculate the flux density using:

$$B = \frac{\Phi}{A}$$

$\left(\right.$Note that the flux required is $\frac{0 \cdot 5}{1000}$ Wb and the cross-sectional area is $\frac{500}{1\,000\,000}$ m².$\left.\right)$

$$B = \frac{0 \cdot 5 \times 1\,000\,000}{1000 \times 500}$$

$$B = 1 \text{ Wb/m}^2 \text{ or } 1 \text{ T}$$

111

The value of magnetizing force needed to produce this flux density in mild steel is given by the magnetization curve (Fig. 6.5) as

$$H = 300 \text{ A t/m}$$

The total m.m.f. required is given by

$$F = H \times l$$

As the length of circuit traversed by the outer lines of magnetic flux is greater than that traversed by the inner lines the correct value to use for "l" is the mean or average length of path.

$$l = \pi d$$

where d is the *mean* diameter

$$l = \pi \times \frac{20}{100}$$

$$l = 0.628 \text{ m}$$

$$F = 300 \times 0.628$$

$$F = 188 \text{ A t}$$

The current is now given by:

$$I = \frac{F}{N}$$

$$I = \frac{188}{500}$$

$$\underline{I = 0.376 \text{ A}}$$

6.5 Permeability

6.51 The ratio of the flux density in a material to the magnetizing force acting on it is called the *absolute permeability* (symbol μ) of the material.

Thus:

$$\mu = \frac{B}{H}$$

This ratio gives a measure of how easily the magnetic flux passes through the material concerned. For magnetic materials the permeability is not constant because the magnetization curve is not a straight line but, for non-magnetic materials such as air, the permeability is constant. The permeability of free space has the value $4\pi \times 10^{-7}$ in the SI system and is denoted by the symbol μ_0. All non-magnetic materials may be assumed to have this value of permeability.

112

6.52 In many cases the permeability of a magnetic material is not quoted as an absolute value, the value relative to that for space being quoted instead. This is called the *relative* permeability (symbol μ_r) and gives a measure of the increase in magnetic flux obtained when a non-magnetic material is replaced by the magnetic material. For purposes of calculation the absolute permeability is required and may be found by using the formula:

$$\mu = \mu_0 \times \mu_r$$

Example 6.4

The magnetic flux in a magnetic circuit follows a path of length 0·4 m in a material having a cross-sectional area of 400 mm² and a relative permeability of 750. What value of m.m.f. is required to produce a flux of 0·32 mWb in this material?

Solution

$$B = \frac{\Phi}{A}$$

$$B = \frac{0·32}{1000} \times \frac{1\,000\,000}{400}$$

$$B = 0·8 \text{ Wb/m}^2 \text{ or } 0·8 \text{ T}$$

$$\mu = \mu_0 \times \mu_r$$

$$\mu = 4\pi \times 10^{-7} \times 750$$

$$H = \frac{B}{\mu}$$

$$H = \frac{0·8}{4\pi \times 10^{-7} \times 750}$$

$$H = 848·6 \text{ A t/m}$$

$$F = H \times l$$

$$F = 848·6 \times 0·4$$

$$\underline{F = 339 \text{ A t}}$$

6.6 Composite Magnetic Circuits

6.61 Practical magnetic circuits are seldom completely uniform throughout their length. The flux may have to pass through various materials including short lengths of air (air gaps) so as to complete its path. In order to calculate the total m.m.f. required to force a given flux through a number of sections in series it is necessary to calculate separately

113

the m.m.f. required for each part of the magnetic circuit in turn. The total m.m.f. is equal to the sum of the individual m.m.f.'s for each part of the circuit.

6.62   For a magnetic circuit having a constant permeability the magnetic flux is directly proportional to the magneto-motive force. This relationship is similar to that between electric current and electro-motive force in a simple electric circuit. The ratio of m.m.f. to flux in a magnetic circuit is called the *reluctance* (symbol S) of the magnetic circuit, in the same way that the ratio of e.m.f. to current is called the resistance of the electric circuit. Thus the reluctance of a section of a magnetic circuit is given by the formula:

$$S = \frac{F}{\Phi}$$

6.63   A formula for the reluctance of a section of magnetic circuit of length, l, absolute permeability, μ, and cross-sectional area, A, can be deduced as follows:

$$S = \frac{F}{\Phi} \qquad \text{(i)}$$

It has already been established that

$$F = IN \quad \text{and} \quad \Phi = BA$$

Substituting these relation for F and Φ in equation (i) gives:

$$S = \frac{IN}{BA} \qquad \text{(ii)}$$

But

$$B = \mu H \quad \text{and} \quad H = \frac{IN}{l}$$

whence

$$B = \frac{\mu IN}{l}$$

Substituting this result for B in equation (ii) gives:

$$S = \frac{IN}{\dfrac{\mu IN \cdot A}{l}} = \frac{INl}{\mu INA}$$

$$\text{i.e.} \quad S = \frac{l}{\mu A}$$

6.64   The table on the facing page compares the relations which hold in magnetic and electric circuits.

Comparison of Electric and Magnetic Circuits	
Electric Circuit 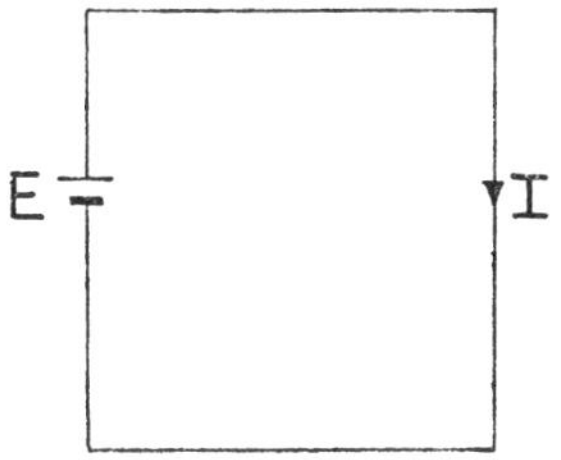	Magnetic Circuit 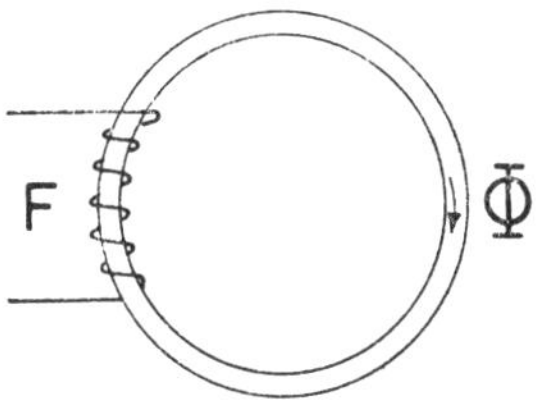

$$R = \frac{E}{I}$$

$$R = \frac{\rho l}{A}$$

$$S = \frac{F}{\Phi}$$

$$S = \frac{l}{\mu A}$$

Example 6.5

A coil of insulated wire of 500 turns and of resistance 4 ohms is closely wound on an iron ring. The ring has a uniform cross-sectional area of 700 mm² and a mean diameter of 0·25 m. Calculate the total flux in the ring when a d.c. supply at 6 V is applied to the ends of the winding. Assume a relative permeability of 550.

Explain the general effect of making a small air gap by cutting the iron ring radially at one point.　　　　　　　　　　　　　　C.G.L.I.

Solution

$$I = \frac{E}{R}$$

$$I = \frac{6}{4}$$

$$I = 1\cdot5 \text{ A}$$

$$H = \frac{IN}{l}$$

$$H = \frac{1\cdot5 \times 500}{\pi \times 0\cdot25}$$

$$H = \frac{3000}{\pi} \text{ A t/m}$$

$$\mu = \mu_0\mu_r$$

$$= 4\pi \times 10^{-7} \times 550$$

$$B = \mu H$$

$$B = 4\pi \times 10^{-7} \times 550 \times \frac{3000}{\pi}$$

(Note that as π was retained in the denominator when calculating H, it now cancels, so simplifying the arithmetic.)

$$B = 660 \times 10^{-3} \text{ Wb/m}^2 \text{ or } 660 \times 10^{-3} \text{ T}$$

$$\Phi = BA$$

$$\Phi = 660 \times 10^{-3} \times \frac{700}{1\,000\,000}$$

$$\underline{\Phi = 0.462 \times 10^{-3} \text{ Wb or } 0.462 \text{ mWb}}$$

As the permeability of air (μ_0) is 550 times lower than that of the iron that it replaces, the reluctance of the magnetic circuit is increased by cutting a slot in it. This gives rise to a reduction in the total flux.

6.7 Electric Fields

6.71 When an insulating rod, such as an ebonite rod, is rubbed it becomes electrified and has the ability to attract light particles such as small pieces of paper. Experiments have shown that all substances can in fact be electrified by friction but conductors have to be well insulated before the effect can be demonstrated, as otherwise the charge produced by the friction of the rubbing leaks away as quickly as it is formed. Charges of this sort are often called *electrostatic* charges, to distinguish them from electric currents which are electricity in motion. Electrostatic charges are due to accumulations (or absences) of electrons, just as electric currents are due to the movement of electrons, and so throughout the remainder of this book they will be referred to simply as *electric charges*.

6.72 In Chapter 1 it was stated that all substances are composed of atoms which contain electrons. Thus any object contains a natural store of electrons. If some extra electrons are added to an object it becomes negatively charged but, if some electrons are removed from an object, it becomes positively charged. When an ebonite rod is rubbed with fur some electrons are transferred from the fur to the ebonite so the ebonite becomes negatively charged and the fur positively charged. If a battery is connected between two insulated metal spheres, as shown in Fig. 6.7, the e.m.f. of the battery forces some electrons from one sphere to the other. These electrons, flowing through the battery constitute a current, I, which flows for a very short time, t, and the amount of charge transferred is $Q = It$ coulombs. When the battery is removed the charge is trapped on the spheres as they are insulated but, if the spheres are subsequently joined by a conductor,

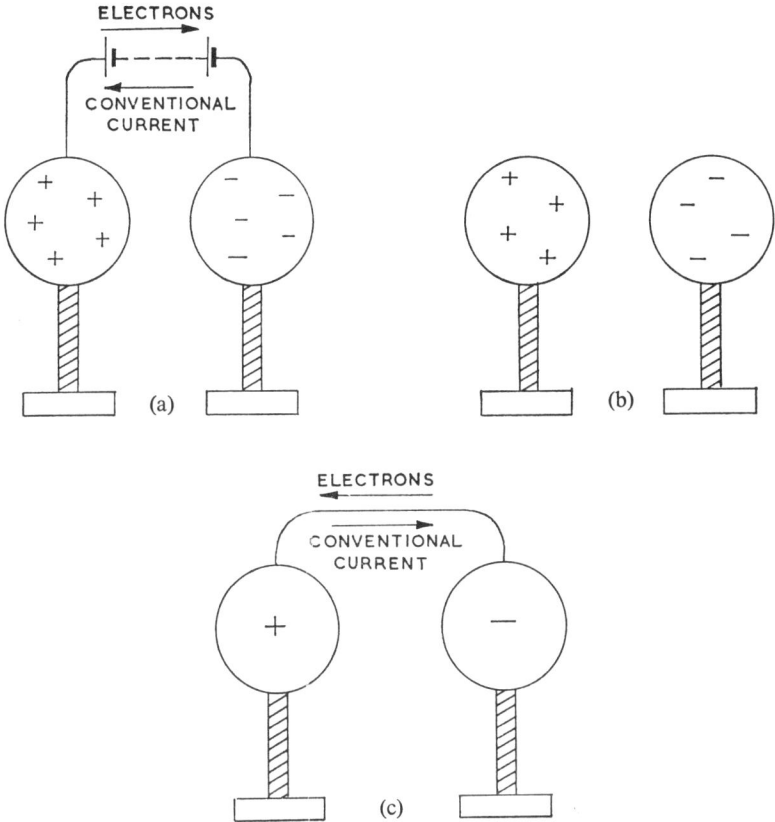

FIGURE 6.7   Electric Charges.
(a) Charging current flows for a short time on applying an e.m.f.
(b) Charge trapped on insulated spheres.
(c) Discharge current flows for a short time when spheres are joined by a conductor.

the same quantity Q of electricity will flow back from one sphere to the other, so discharging them.

6.73 Experiments on electrically charged bodies show that bodies with like charges repel each other while bodies with unlike charges attract each other. If a small light body with a positive charge is placed near to another charged object, it will experience a force. The region in which this force due to a charged object can be detected is called an *electric field*. Electric fields can be mapped by drawing lines which show the direction in which a small positively charged particle would move under the influence of the field. Figure 6.8 shows some typical electric fields.

117

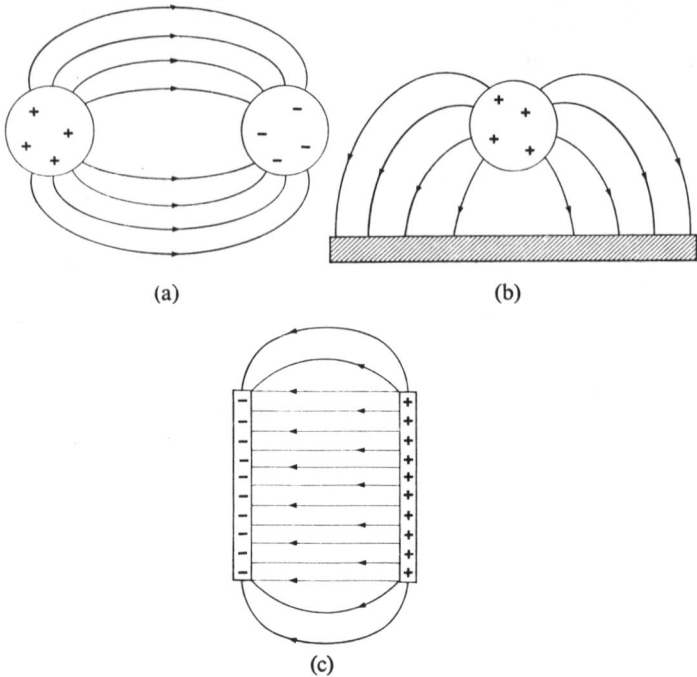

(a)

(b)

(c)

FIGURE 6.8   Electric Fields.
(a) Field between two oppositely charged spheres.
(b) Field between one charged sphere and earth.
(c) Field between two oppositely charged flat parallel metal plates.

6.8   Capacitance

6.81   In paragraph 6.72 it was shown that two objects, insulated from each other, could be used to store an electric charge. Any arrangement of this sort, which can store electric charges, is said to possess *capacitance*. The storage capacity of any arrangement is measured by determining the amount of charge which is stored when there is a p.d. of one volt between the two objects. An arrangement which can store a charge of one coulomb for a p.d. of one volt is said to have a capacitance of one **farad** (abbreviation F, symbol C). In practice this proves to be a very large capacitance and so a smaller unit, the microfarad (abbreviation μF) is commonly used to measure capacitances.

$$1 \ \mu\text{F} = 10^{-6} \ \text{F}.$$

6.82   A *capacitor* is a piece of apparatus which is designed to store electrical energy in the form of an electric charge. It usually consists of two conducting surfaces known as *plates* or *electrodes*. These plates are

118

arranged close to each other, but not touching, and are separated by an insulating material which is called the *dielectric*. The capacitance of a capacitor depends on three main factors:

(a) The area of the plates. (The greater the area the greater the capacitance.)
(b) The spacing between the plates. (The closer the spacing the greater the capacitance.)
(c) The nature of the dielectric.

Figure 6.9(a) shows a simple capacitor consisting of two metal plates separated by a small air gap. In this case the air between the plates forms the dielectric. Figure 6.9(b) shows the theoretical symbol for representing a capacitor in a circuit diagram. Figure 6.9(c) shows a rolled metal foil capacitor using paper as a dielectric. This type of capacitor uses aluminium foil sheets as the plates, the foil sheets being separated by a layer of thin, impregnated paper which serves as the dielectric. As the foil sheets and the paper are both flexible the capacitor can be rolled up to a convenient form, and is then often enclosed in a plastic, cardboard or metal case. Although the use of paper as the dielectric is very common, this type of capacitor can also be made using polythene or similar flexible plastic insulation as the dielectric. Large paper dielectric capacitors are often immersed in oil to improve the insulation, keep moisture at bay and assist cooling. *Electrolytic* capacitors are somewhat similar in construction

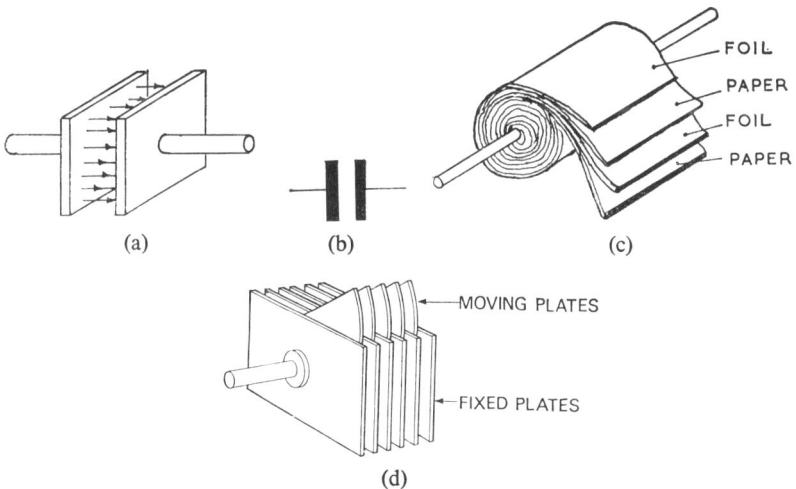

FIGURE 6.9   Capacitors.
(a) Parallel plate capacitor.
(b) Theoretical symbol for capacitor.
(c) Rolled foil (paper) capacitor.
(d) Variable capacitor.

to paper capacitors but, in this case, the paper used is an absorbent type impregnated with a solution containing borax. When an initial forming current is passed through the capacitor by the manufacturers, electrolysis takes place and an oxide skin is formed on the positive aluminium foil (anode) and this acts as the dielectric of the capacitor. Electrolytic capacitors provide a large capacitance for a small physical size, but have the disadvantage that, if a voltage of the wrong polarity is applied to them, the oxide skin is damaged and the capacitor destroyed. Thus the ordinary type of electrolytic capacitor cannot be used in alternating current applications, although special types for a.c. use are available.

A variable capacitor can be produced by arranging that one set of capacitor plates can be moved relative to the other set. In this way, either the spacing or the facing area between the plates can be varied, so altering the capacitance. Figure 6.9(d) shows a variable capacitor in which one set of plates can be moved so as to alter the effective area between them.

6.83   The amount of charge stored by a capacitor depends both on the capacitance and on the p.d. between the plates.

i.e. $$Q \text{ (coulombs)} = V \text{ (volts)} \times C \text{ (farads)}$$

Example 6.6

A 200 μF capacitor is charged by a steady current of 0·2 A flowing for 3 seconds. What is the final value of the charge stored in the capacitor and the p.d. between its plates? If the capacitor is subsequently discharged in 60 microseconds what is the average discharge current?

Solution

$$\text{Charge stored} = It$$

$$Q = 0·2 \times 3$$

$$\underline{Q = 0·6 \text{ coulomb}}$$

$$Q = VC$$

p.d. between plates, $$V = \frac{Q}{C}$$

$$V = \frac{0·6}{200 \times 10^{-6}}$$

$$V = \frac{0·6 \times 1\,000\,000}{200}$$

$$\underline{V = 3000 \text{ V}}$$

Discharge current,

$$I = \frac{Q}{t}$$

$$I = \frac{0{\cdot}6}{60 \times 10^{-6}}$$

(Note, one microsecond $= 10^{-6}$ seconds.)

$$I = \frac{0{\cdot}6 \times 1\,000\,000}{60}$$

$$\underline{I = 10\,000 \text{ A}}$$

This example illustrates one use of a capacitor, namely that it can be charged comparatively slowly and then used to provide a heavy current lasting for a very brief instant.

6.84 As a capacitor can supply both voltage and current for a short period it may be regarded as a store of energy. Consider a capacitor storing a charge Q at a voltage V. The capacitor could supply a current I for t seconds, where $Q = It$. During this time the voltage decreases from V to 0 volts, so the average voltage is $V/2$. The current I is given by $I = Q/t$

$$\therefore \quad I = \frac{VC}{t}$$

it follows that the energy is given by:

$$\text{joules} = \text{current} \times \text{average voltage} \times \text{time}$$

$$W = \frac{VC}{t} \times \frac{V}{2} \times t$$

$$W = \tfrac{1}{2}CV^2$$

6.85 In practice when a capacitor discharges through a resistor the current does not remain constant but varies as shown in Fig. 6.10(a). This is because as the charge flows out from the capacitor its voltage is correspondingly reduced, as shown in Fig. 6.10(b), and this in turn reduces the value of the current. The initial value of the current flowing is given by $I = V/R$ (where V is the initial voltage of the capacitor). If this current were maintained, the capacitor would discharge in a time given by:

$$t = \frac{Q}{I}$$

$$t = \frac{VC}{V/R}$$

$$t = CR \text{ seconds}$$

This time is called the *time constant* (symbol τ) of the capacitor and resistance combination, whence $\tau = CR$.

It can be shown that in practice the capacitor is only approximately $\frac{2}{3}$ discharged in a period equal to one time constant. It loses $\frac{2}{3}$ of its remaining charge in the next period of one time constant and so on. Figures 6.10(c) and (d) show what happens when a capacitor is charged; again the action is not instantaneous, charging being only $\frac{2}{3}$ completed in a period of one time constant.

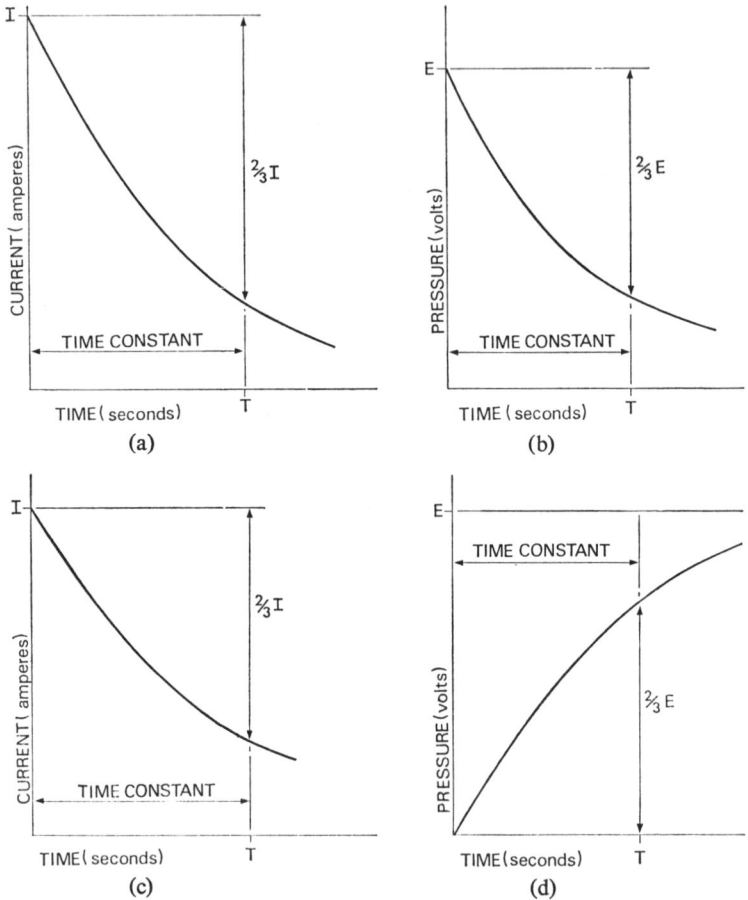

FIGURE 6.10 Capacitor charge and discharge curves.
(a) Current during discharge. (c) Current during charge.
(b) Voltage during discharge. (d) Voltage during charge.

6.9 Combination of Capacitors

6.91 Figure 6.11 shows three capacitors connected in series to a source of e.m.f. (E). When the e.m.f. is first applied a current I flows for t

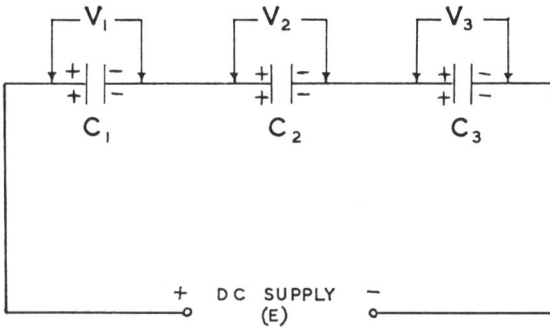

FIGURE 6.11   Capacitors in Series.

seconds charging the capacitors and as each capacitor is charged by the same current for the same time, each must acquire the same amount of charge, Q.

As $Q = VC$ it follows that

$$\frac{V_1}{Q} = \frac{1}{C_1}$$

$$\frac{V_2}{Q} = \frac{1}{C_2}$$

and

$$\frac{V_3}{Q} = \frac{1}{C_3}$$

Also, since V_1, V_2 and V_3 are in series:

$$E = V_1 + V_2 + V_3$$

The total capacitance of the arrangement, C_T is given by

$$C_T = \frac{Q}{E}$$

$$\frac{1}{C_T} = \frac{E}{Q}$$

$$\frac{1}{C_T} = \frac{V_1 + V_2 + V_3}{Q}$$

$$\frac{1}{C_T} = \frac{V_1}{Q} + \frac{V_2}{Q} + \frac{V_3}{Q}$$

$$\frac{1}{C_T} = \frac{1}{C_1} + \frac{1}{C_2} + \frac{1}{C_3}$$

123

In general, for any number of capacitors joined in series:

$$\frac{1}{C_T} = \frac{1}{C_1} + \frac{1}{C_2} + \cdots \frac{1}{C_n}$$

6.92 Figure 6.12 shows three capacitors connected in parallel with a source of e.m.f. (E). In this case each capacitor charges separately until the p.d. between its plates is equal to E volts. Hence the charge in C_1 is given by the formula:

$$Q_1 = EC_1$$

$$C_1 = \frac{Q_1}{E}$$

Similarly $\quad C_2 = \frac{Q_2}{E}$

and $\quad C_3 = \frac{Q_3}{E}$

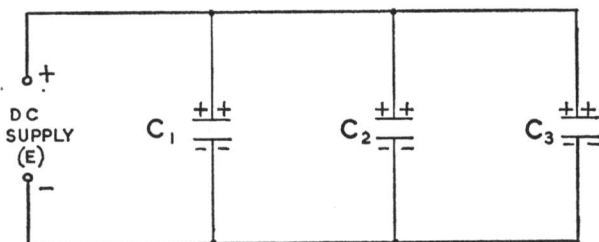

FIGURE 6.12   Capacitors in Parallel.

As each capacitor has charged separately, i.e. each capacitor is initially charged by its own current, the total charge is:

$$Q_T = Q_1 + Q_2 + Q_3$$

$$C_T = \frac{Q_T}{E}$$

$$C_T = \frac{Q_1}{E} + \frac{Q_2}{E} + \frac{Q_3}{E}$$

$$C_T = C_1 + C_2 + C_3$$

In general, for any number of capacitors connected in parallel:

$$C_T = C_1 + C_2 + \cdots + C_n$$

Example 6.7

(a) Describe, with sketches, the construction of a small paper capacitor. Give one instance of the use of such a capacitor and state its function in the particular circuit.

(b) Three capacitors of values 6 μF, 8 μF and 10 μF respectively are connected (i) in series and (ii) in parallel. Calculate the resultant capacitance in each case.
C.G.L.I.

Solution

(a) A paper insulated capacitor is described in paragraph 6.82 and illustrated in Fig. 6.9(c).

Capacitors are widely used in a.c. circuits a typical example being the use of capacitors for power factor correction which is fully described in Chapter 9.

(b) (i) For capacitors in series:

$$\frac{1}{C_T} = \frac{1}{C_1} + \frac{1}{C_2} + \frac{1}{C_3}$$

$$\frac{1}{C_T} = \frac{1}{6} + \frac{1}{8} + \frac{1}{10}$$

$$\frac{1}{C_T} = \frac{20 + 15 + 12}{120}$$

$$\frac{1}{C_T} = \frac{47}{120}$$

$$C_T = \frac{120}{47}$$

$$\underline{C_T = 2\cdot55\ \mu F}$$

(ii) For capacitors in parallel:

$$C_T = C_1 + C_2 + C_3$$
$$C_T = 6 + 8 + 10$$
$$\underline{C_T = 24\ \mu F}$$

SUMMARY OF FORMULAE

Magnetic flux density: $\quad B = \dfrac{\Phi}{A}$

$$B = \mu H$$

Magneto motive force: $\quad F = IN$

Magnetizing force⟍ or magnetic field strength:	$H = \dfrac{IN}{l}$
Permeability:	$\mu = \mu_0\mu_r$
	$\mu_0 = 4\pi \times 10^{-7}$
Reluctance:	$S = \dfrac{F}{\Phi}$
	$S = \dfrac{l}{\mu A}$
Charge in capacitor:	$Q = VC$
Energy stored in capacitor:	$W = \tfrac{1}{2}CV^2$
Time constant of CR combination:	$\tau = CR$
Capacitors in series:	$\dfrac{1}{C_T} = \dfrac{1}{C_1} + \dfrac{1}{C_2} + \cdots \dfrac{1}{C_n}$
Capacitors in parallel:	$C_T = C_1 + C_2 + \cdots C_n$

EXERCISE 6

1. A magnetic circuit has a cross-sectional area of 0.6 m². Calculate the flux density when the total flux is 0.5 Wb. Determine the magnetizing force required if the material from which the circuit is made is mild steel. (Use the B/H curves of Fig. 6.5.)

2. Sketch the shape of the magnetic field produced (a) by a bar magnet (b) by a solenoid carrying an electric current; carefully indicating the direction of the current and the polarity of the magnetic field. What properties are desirable in the material from which a permanent magnet is to be made.

3. Explain carefully the following terms used in connection with a magnetic field;
   (a) Saturation.
   (b) Hysteresis.
   (c) Residual magnetism
   (d) Permeability.

4. A coil of insulated wire of 400 turns and of resistance 0·25 ohm, is wound tightly round an iron ring and is connected to a d.c. supply of 4 volts. The iron ring is of uniform cross-sectional area 600 mm², and of mean diameter 150 mm. The permeability of the ring may be taken as 450. Calculate the total flux in the iron. What would be the effect of an air gap in the ring? C.G.L.I.

5. Sketch a typical magnetization curve of a sample of magnetic iron. An iron ring of mean diameter 0·26 m and of cross-sectional area 800 mm² is closely wound with 550 turns of insulated wire of resistance 35 ohms. Calculate the magnetic flux in the ring when a d.c. supply of 210 V is applied to the ends of the winding. Assume a relative permeability of 490. ($\mu_0 = 4\pi \times 10^{-7}$ in the SI system.) C.G.L.I.

6. A wrought iron ring of mean diameter 0·3 m and cross-sectional area 1000 mm² is wound with a coil of 500 turns. What current in the coil will produce a flux of 1·5 mWb in the ring? Calculate the relative permeability of the iron for this value of flux. (Use the B/H curves of Fig. 6.5.)

7. A magnetic circuit made of cast steel has a length of 1·5 m and a cross-sectional area of 0·2 m². What value of m.m.f. is required to produce a flux of 0.3 Wb in this magnetic circuit? If a slot is cut in the circuit so increasing its reluctance by 50%, what new value of m.m.f. would be required to maintain the flux at its original value? (Use the B/H curves of Fig. 6.5.)

8. A coil of insulated wire of 600 turns and of resistance 4·5 ohms is closely wound on a uniform iron ring. The ring has a mean diameter of 0·3 m, and a cross-sectional area of 800 mm². Calculate the total flux in the ring when a d.c. supply at 27 V is applied to the ends of the coil. Assume a relative permeability of 480.

    Sketch a typical magnetization curve of a sample of magnetic iron.   C.G.L.I.

9. An iron ring of mean diameter 180 mm and uniform cross-sectional area 700 mm², is wound with 650 turns of insulated wire of total resistance 0·5 ohm. Calculate the total flux in the ring when a d.c. supply of 4 V is applied to the ends of the coil. Assume a relative permeability of 520.

    Describe and explain the general effect obtained by making a thin radial cut at one point in the ring. ($\mu_0 = 4\pi \times 10^{-7}$ in the SI system.)   C.G.L.I.

10. What is meant by the term *capacitance*?

    Three capacitors of values 8 μF, 12 μF and 24 μF are joined together
    (a) in series
    (b) in parallel
    (c) so that the 8 μF capacitor is connected in series with the other capacitors in parallel
    Calculate the total capacitance in each case.

11. Upon what three factors does the capitance of a capacitor depend?

    A 100 μF capacitor, which is initially completely discharged, is charged by a steady current of 0·01 mA flowing for 2 seconds. What are the values of the final charge in the capacitor and the p.d. between its plates. If the capacitor is subsequently discharged by a steady current flowing for 0·01 seconds, what is the value of the current?

12. Describe with sketches the construction of a small paper capacitor. What is meant by the term *dielectric*?

    A capacitor is charged until the p.d. between its plates is 500 V and the total charge in the capacitor is then 1000 microcoulombs. What is the value of this capacitor?

    If a second capacitor of value 4 μF were connected in series with the above capacitor what would then be the total capacitance?

7 Electro-Magnetic Effects

7.1 Electromagnets

7.11 An electromagnet consists of a coil of insulated wire wound on a suitable soft iron core. When an electric current is passed through the coil, the core becomes strongly magnetized. The core is magnetically soft so that when the current is switched off it retains hardly any magnetism and no longer exerts any magnetic attraction. Electromagnets are often

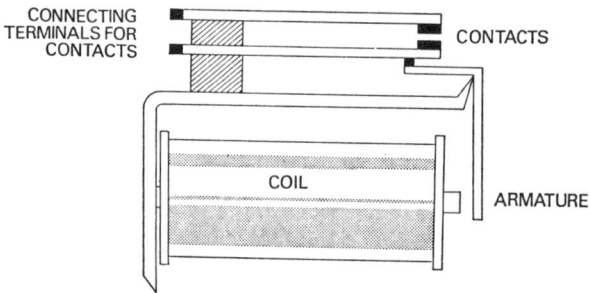

FIGURE 7.1   Relay.

used instead of hooks on cranes for lifting magnetic materials, such as scrap iron and sheet steel; they are also used in devices such as electric bells and relays which are described below. The poles of electric generators and motors are often electromagnets.

7.12 A *relay* is an electromagnetically operated switch. Figure 7.1 shows one form of relay. When current flows through the coil the armature is attracted and operates the contacts. Although the type of relay illustrated in Figure 7.1 has contacts which are capable of carrying only light currents, magnetically operated switches can be designed with heavy duty contacts. Such switches are often called "contactors" and they are widely used for such purposes as motor starting and remote switching of mains supplies.

7.13 The trembler bell as shown in Fig. 7.2 provides yet another example of the use of an electromagnet. When an electric current flows through the coils the armature is attracted by the electromagnet so

128

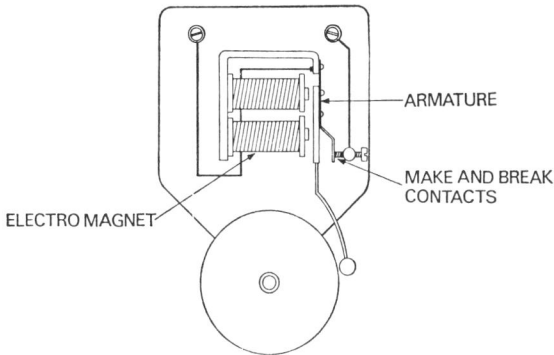

FIGURE 7.2   Trembler Bell.

causing the hammer to strike the gong. The movement of the armature opens the make and break contacts so interrupting the current and switching off the electromagnet. This allows the armature to return under the influence of the spring so closing the contacts again and allowing the cycle to repeat itself over and over again. The armature continues to "tremble," ringing the bell until the supply is switched off. The mechanism of a buzzer is similar to that of a trembler bell with the striker and gong omitted.

7.2   Force on a Current-Carrying Conductor

7.21   Whenever a conductor, situated in a magnetic field, carries an electric current, it experiences a force. The direction in which this force

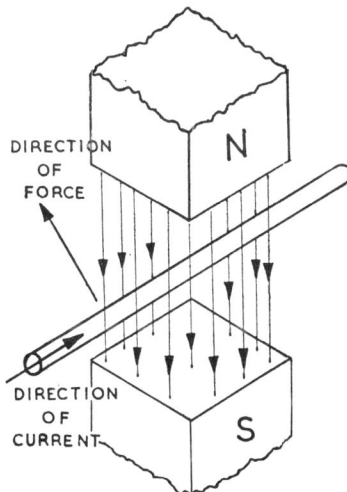

FIGURE 7.3   Left Hand Rule.

acts can be found by using the *left hand rule* as follows: Hold the left hand flat so as to receive the magnetic flux on the palm, with the fingers pointing in the direction of the current and the thumb extended. Then the thumb points in the direction that the force acts.

7.22 The magnitude of the force produced by a current is proportional to the strength of the magnetic field, (B) the value of the current (I) and the active length of the conductor (l), i.e. the length of conductor actually affected by the magnetic field. The magnitude of the force is given by:

$$F = BIl \text{ newtons}$$

7.23 A simple example of the application of this effect is the moving coil loudspeaker shown in Fig. 7.4. When an electric current flows through the coil it produces a force which moves the coil of the loudspeaker backwards or forwards according to the direction of the current.

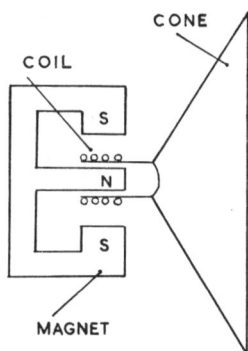

FIGURE 7.4 Moving Coil Loudspeaker.

Example 7.1

The coil of a moving coil loudspeaker has 20 turns of 15 mm diameter, situated in a magnetic field of strength 0·8 T. What force is produced when a current of 0·1 A flows through the coil?

Solution

Active length of conductor,

$$l = 20 \times \pi \times \frac{15}{1000}$$

$$l = 0\cdot943 \text{ m}$$

$$F = BIl$$

$$F = 0\cdot8 \times 0\cdot1 \times 0\cdot943$$

$$\underline{F = 0\cdot0754\,\text{N}}$$

7.3 Torque Produced by Electric Currents

7.31 If a coil is situated in a magnetic field and an electric current is passed through it, forces act on the sides of the coil as shown in Fig. 7.5. These forces act in opposite directions and so tend to make the coil rotate, i.e. they produce a torque.

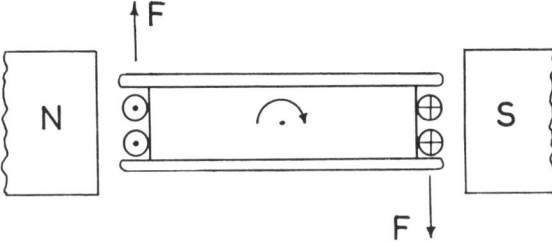

FIGURE 7.5   Torque Produced by Coil.

7.32 For a rectangular coil, as shown in Fig. 7.6, each conductor in the coil side experiences a force given by:

$$F = BIl$$

As there are N conductors the total force on each coil side is

$$F = BIlN$$

and as there are two coil sides each at a radius $d/2$, the total torque will be

$$T = 2BIlN \times \frac{d}{2} \text{ newton metre}$$

$$T = BINld \text{ N m}$$

or $\qquad\qquad T = BINA \text{ N m}$

where $A =$ area enclosed by coil.

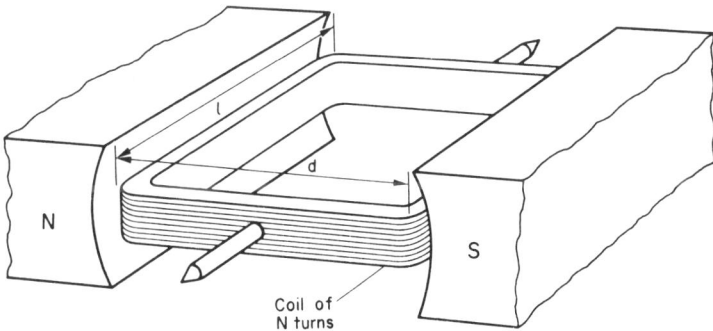

FIGURE 7.6   Rectangular Coil in Magnetic Field.

131

7.33 A simple example of the application of this effect is the moving coil meter. (See Chapter 11.)

Example 7.2

The coil of a moving coil meter has 20 turns and is wound on a rectangular former of dimensions 20 mm × 30 mm, pivoted at the centre of the shorter side. The coil is situated in a magnetic field of strength 0·9 T. What torque is produced when a current of 2 mA flows through the coil?

Solution

$$T = BINld$$
$$T = 0.9 \times \frac{2}{1000} \times 20 \times \frac{30}{1000} \times \frac{20}{1000}$$
$$T = 0.000\ 021\ 6\ \text{N m}$$
$$\underline{T = 0.0216\ \text{mN m}}$$

7.4 Induced E.M.Fs.

7.41 Whenever a conductor moves through a magnetic field an e.m.f. is induced in it. This effect can easily be demonstrated by connecting a coil to a sensitive milli-voltmeter as shown in Fig. 7.7 and moving a magnet in the coil. It is found that:

(a) An e.m.f. is induced if either the coil or the magnet is moved relative to the other.
(b) The polarity of the induced e.m.f. depends on the polarity of the magnet and on the direction of motion.
(c) The magnitude of the induced e.m.f. depends on the rate at which the magnet moves relative to the coil.

7.42 The direction in which the induced e.m.f. acts can be found by using the *right hand rule* as follows: Hold the right hand flat so as to

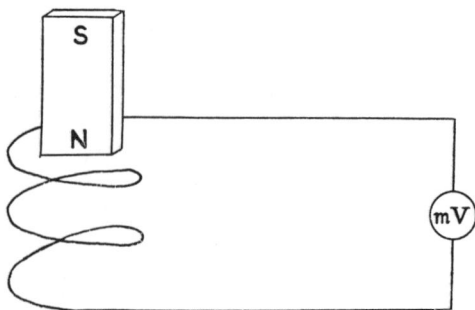

FIGURE 7.7 Demonstration of Induced Voltages.

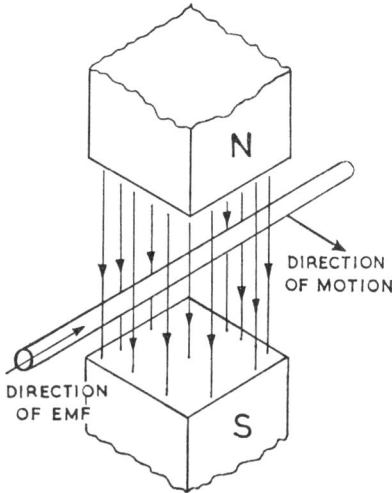

FIGURE 7.8 Direction of Induced e.m.f.

receive the magnetic flux on the palm, with the thumb extended and pointed in the direction that the conductor moves relative to the field. Then the fingers point in the direction of the induced e.m.f.

7.43 If a circuit is completed so that the induced e.m.f. can produce an electric current, this current will react with the magnetic field so producing a force. By applying the left hand rule it can be seen that the direction of this force opposes the motion of the conductor. This result is known as *Lenz's law* which can be stated as: *any induced current tends to oppose the motion producing it.*

7.5 Magnitude of Induced E.M.Fs.

7.51 The magnitude of an induced e.m.f. is equal to the rate at which the conductor passes through the magnetic flux. If a conductor passes steadily through Φ webers of magnetic flux every second, then Φ volts are induced in it.

$$e \text{ (induced e.m.f.)} = \frac{\Phi \text{ (flux passed through)}}{t \text{ (time taken to pass through the flux)}}$$

7.52 If a conductor l metres long moves at right angles to a magnetic field at a velocity of v metres/second then it sweeps an area of lv square metres every second as shown in Fig. 7.9. The flux (Φ) passed through by the conductor is the product of the area and flux density,

i.e. $$\Phi = BA$$

The amount of flux cut per second $= Blv$

133

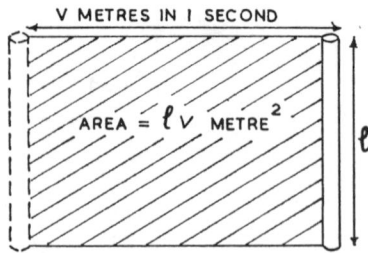

FIGURE 7.9   Area swept by Moving Conductor.

But, as the e.m.f. (*e*) induced equals the amount of flux passed through in one second, it follows that:

$$e = Blv$$

Example 7.3

A conductor 0·5 m long moves at a velocity of 15 m/s through a magnetic field of strength 1·1 T, the conductor and its direction of motion both being at right angles to the magnetic field. What value of e.m.f. is induced in the conductor?

Solution

$$e = Blv$$
$$e = 1·1 \times 0·5 \times 15$$
$$e = 8·25 \text{ V}$$

7.53   Whenever the magnetic flux linked with a coil changes there is an e.m.f. induced in the coil. Figure 7.10 shows a coil of N turns linked with Φ_1 webers of magnetic flux. If the flux is increased to Φ_2 webers in t seconds, then a flux of $\Phi_2 - \Phi_1$ webers has entered the coil. In passing through one turn of the coil this must induce an e.m.f. of

$$\frac{\Phi_2 - \Phi_1}{t} \text{ volts}$$

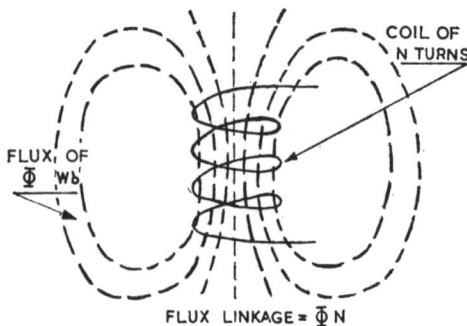

FIGURE 7.10   Flux Linkage.

and hence, for the whole coil of N turns, the induced e.m.f. is given by:

$$e = \frac{(\Phi_2 - \Phi_1)N}{t}$$

The product of magnetic flux and turns, ΦN, is often called the *flux linkage* of the coil and the above formula shows that the induced e.m.f. is equal to the rate of change of flux linkage.

Thus:
$$e = \frac{\text{Change in flux linkage}}{\text{Time taken (seconds)}}$$

7.6 Inductance

7.61 The magnetic flux linked with a coil may well be due to an electric current flowing in the coil itself. Whenever this current changes there must be a corresponding change in the magnetic flux, and so an e.m.f. will be induced in the coil. This e.m.f., which exists only while the current is changing, is called the *e.m.f. of self induction*. Applying the principle of Lenz's law shows that a self induced e.m.f. must always oppose the change in current which produces it, i.e. if the current tends to RISE the induced e.m.f. acts AGAINST the direction of the current, while if the current tends to FALL the induced e.m.f. acts in the SAME DIRECTION as the current. The ability of a coil to generate self induced voltages is called its *self inductance*. Any means of increasing the magnetic field of the coil, such as winding it on an iron core, will increase its self inductance.

7.62 The self inductance of a coil is measured in **henrys** (symbol L, abbreviation H). A coil has a self inductance of one henry if an e.m.f. of one volt is induced in it by a current changing at a rate of one ampere per second. The e.m.f. induced at any instant in a coil by an increasing current is given by:

$$e = -L \times \text{rate of increase of current}$$

or
$$e = -L \times \frac{\text{Increase in current}}{\text{Time taken for current to increase}}$$

It should be noted that the minus sign in the above expression indicates that the self induced e.m.f. opposes the increase in the current.

Example 7.4

An inductor has a self inductance of 25 H and carries a current of 1·2 A. What e.m.f. is induced in this inductor if the current is interrupted in 0·01 seconds?

Solution

$$e = -L \times \frac{\text{Increase in current}}{\text{Time taken to increase}}$$

Note that in this case the current is reduced from 1·2 A to zero, and that this is a *negative* increase.

$$e = -25 \times \frac{-1·2}{0·01}$$

$$e = +3000 \text{ V}$$

The positive sign of the answer indicates that this e.m.f. acts in a direction which tends to keep the current flowing. This example illustrates the way in which a very high e.m.f. may be momentarily induced when the current in an inductive circuit is suddenly interrupted.

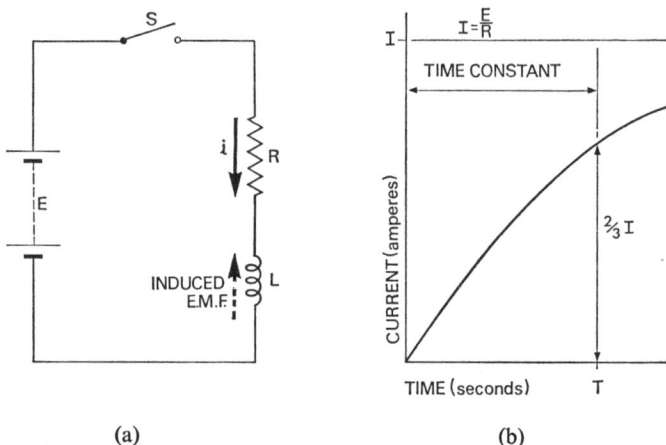

FIGURE 7.11   Rise of Current in an Inductive Circuit.
     (a) Inductive Circuit.
     (b) Rise of Current after Closing Switch.

7.63   Figure 7.11(a) shows a circuit containing resistance and inductance in series with a d.c. supply. When the switch *S* is closed, the current rises from zero to the value required by Ohm's law ($I = E/R$), but while it is rising the self inductance causes an e.m.f. to be induced which opposes the build up of current. Thus the current does not immediately attain its final value but rises gradually as shown in Fig. 7.11(b). The shape of this graph is known as an *exponential* curve. The current rises to approximately $\frac{2}{3}$ (actually 62·8%) of its final value in a time of $\frac{L \text{ (henrys)}}{R \text{ (ohms)}}$ seconds. It then

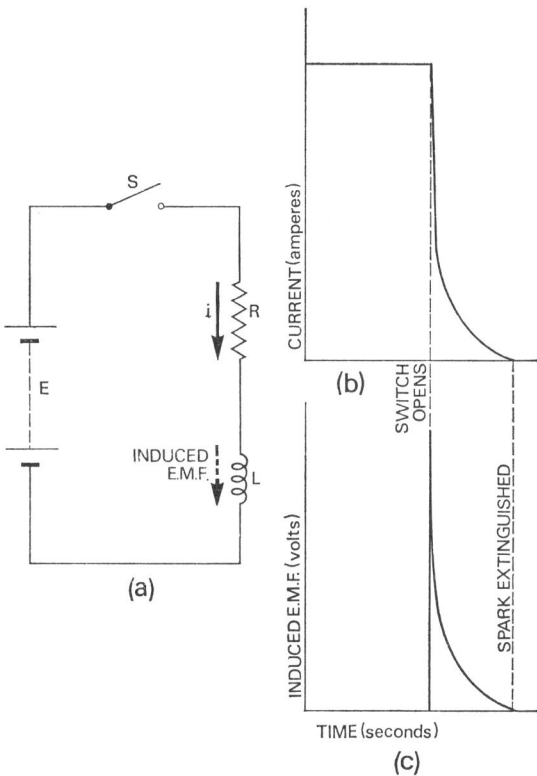

FIGURE 7.12 Interruption of Current in Inductive Circuit.
        (a) Inductive Circuit.
        (b) Current/time Curve
        (c) E.M.F. Induced on Opening Switch.

rises by $\frac{2}{3}$ of the remaining distance in the next L/R seconds, and so on, as shown in Fig. 7.11(b). The time, L/R seconds is called the *time constant* (symbol τ) of the circuit.

$$\tau = \frac{L}{R} \text{ seconds}$$

Figure 7.12 shows what happens when the switch S is suddenly opened. The current falls rapidly, although not instantaneously to zero, and this rapid fall of current causes a high induced e.m.f. This e.m.f. will, in practice, tend to cause a severe spark or arc between the switch contacts as they separate.

7.64 A current-carrying inductor is a store of energy, the energy being stored in the form of a magnetic field. It can be shown that the amount of

137

energy stored in an inductor is:

$$W = \tfrac{1}{2}LI^2$$

When the current flowing through an inductor is interrupted, this energy must somehow be dissipated. The e.m.f. induced on breaking the circuit will, in practice, tend to cause a severe spark or arc between the switch contacts as they separate, and most of the energy stored in the inductor is dissipated in the form of heat in the spark. The remaining energy in the inductor is used in such ways as heating the circuit conductors: some may even be radiated as an electromagnetic wave (radio wave) causing interference with any nearby radio equipment.

7.65   If two coils are arranged so that the magnetic flux produced by one coil links with the turns of the other, as shown in Fig. 7.13, then any change of current in one coil will induce an e.m.f. in the other. An e.m.f. produced in this way is called a mutually induced e.m.f. and the coils are said to possess *mutual inductance* (symbol *M*). Like self inductance, mutual inductance is measured in henrys; the value of the mutually induced e.m.f. being given by:

e.m.f. induced in coil $B = M \times$ rate of change of current in A.

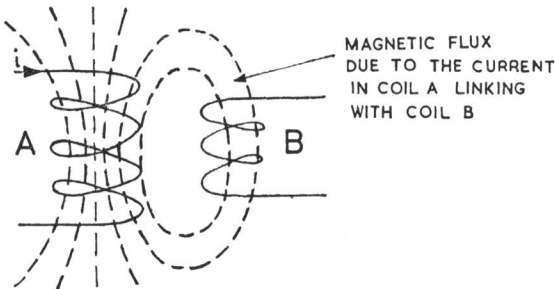

FIGURE 7.13   Mutual Inductance.

7.66   Inductance is often undesirable, for example in a.c. circuits it increases the voltage drop above the value due to resistance alone. For this reason when installing wiring in conduit, the phase and neutral conductors of any circuit should be enclosed in the same conduit tube. In this way the magnetic effects of the currents, which flow in opposite directions in the two conductors, tend to cancel, so keeping the inductance of the circuit to a minimum. Similarly resistors can be non-inductively wound, one method being to arrange the winding in the form of two coils wound in opposite directions so that once again the magnetic effects of the currents tend to cancel.

138

7.7 The Simple Generator

7.71 Figure 7.14(a) shows a single loop coil which rotates between the poles of a magnet. In order to make electrical connection with the coil, use is made of *slip-rings*; that is each of the leads from the coil is connected to a brass or copper ring which rotates with the coil, and contact is made by means of a *brush* which is a rubbing contact pressing on the ring. As the loop rotates, its sides cut through the magnetic flux set up by the

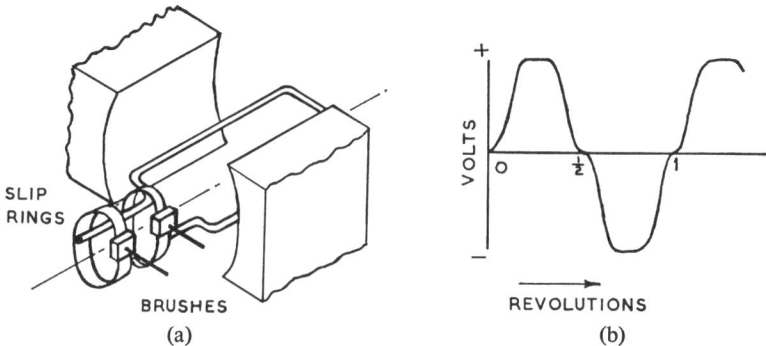

FIGURE 7.14   Simple a.c. Generator.
(a) Single Loop Alternator.
(b) Waveform of Induced e.m.f.

poles, and so an e.m.f. is induced in the loop. This e.m.f. reverses its polarity every half revolution as shown by the graph or waveform illustrated in Fig. 7.14(b). Thus the simple loop generator, with slip-rings, is a form of a.c. generator or alternator. The simple alternator generates one *cycle* of its waveform for every complete revolution that it makes. Nevertheless it is possible to add additional magnetic poles in pairs to the generator, the polarities forming a sequence N, S, N, S, and so on. The alternator will now generate one cycle for every pair of poles during one revolution. So if there are p pairs of poles the machine will generate p cycles per revolution. The *frequency* (symbol f) of an alternator is the number of cycles it generates every second. Frequency is measured in **hertz** (abbreviation Hz). It can be seen that a machine having p pairs of poles and running at N rev/min generates a frequency of:

$$f = \frac{pN}{60} \text{ Hz}$$

7.72 Instead of using two slip-rings to make connection to the rotating coil, it is possible to use one ring which is split into two halves insulated from each other as shown in Fig. 7.15(a). Such an arrangement is called a *commutator*. The function of the commutator is illustrated by Figs.

139

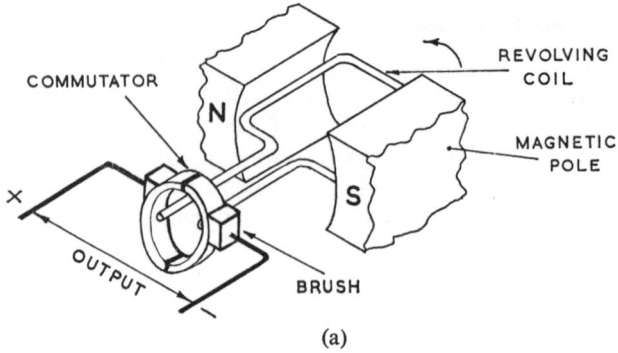

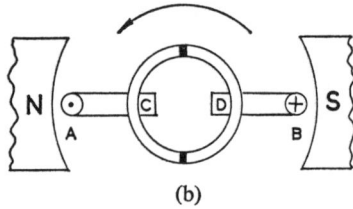

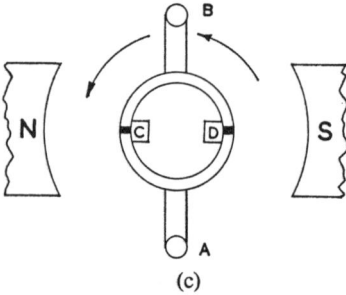

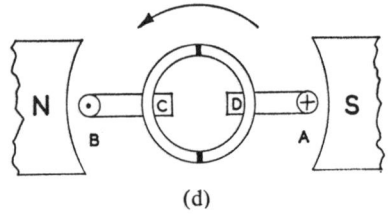

FIGURE 7.15   Simple d.c. Generator.
(a) Single Loop Coil and Commutator.
(b) Coil Generating with Side *A* Under North Pole.
(c) Coil Sides Between Poles and Not Generating.
(d) Coil Generating with Side *A* Under South Pole.   Brushes Still Have Initial Polarity.

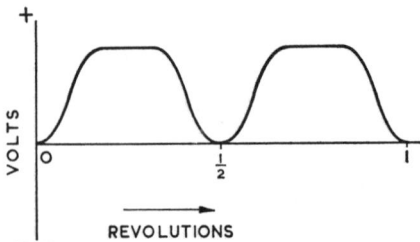

FIGURE 7.16   Waveform of Simple d.c. Generator.

7.15(b), (c) and (d). In Fig. 7.15(b) the direction of the e.m.fs. induced in the coil sides *A* and *B* make brush *C* positive and brush *D* negative in polarity as shown, in accordance with the right hand rule. In Fig. 7.15(c) the coil has advanced by a quarter revolution, neither coil side has an induced e.m.f. as it is moving parallel to the lines of the magnetic field and so is not cutting any magnetic flux. In Fig. 7.15(d) the coil has made a further quarter revolution, the e.m.f.s induced in coil sides *A* and *B* are now in the opposite direction to those previously induced but, as the connection to the coil is made via the commutator, brush *C* still has a positive polarity and brush *D* a negative polarity. It can be seen that the commutator automatically reverses the connections between the rotating coil and the brushes at the same instant that the polarity of the e.m.f. induced in the coil itself reverses. This results in the output from the brushes always being of the same polarity as shown in Fig. 7.16. Thus a simple loop generator, fitted with a commutator, becomes a d.c. generator or dynamo. The e.m.f. produced by any generator depends on the rate at which the active conductors cut through the magnetic field. It follows that the e.m.f. is directly proportional both to the speed at which the machine is driven and to the strength of the magnetic field.

7.8 The Practical D.C. Generator

7.81 A practical d.c. generator consists of three main parts, the armature, commutator and field system. The armature is the revolving part of the machine and usually consists of a laminated iron cylinder with slots equally spaced around its circumference. Coils are fitted into the slots to form the active winding in which the required induced e.m.fs. are developed, a large number of coils being used so that their e.m.f. waves overlap and so give a smoother output than that obtained from one coil acting alone. The commutator forms the link by which electrical connection between the fixed and rotating parts is obtained; it also serves as the means for converting the alternating e.m.f. induced in the armature coils into the required d.c. output. The field system is required to produce the powerful magnetic field which is needed to induce useful e.m.fs. in the armature coils. It is normally made in the form of an electro-magnet with one or more pairs of poles, these poles being magnetised by the action of the *field current* flowing in field coils mounted on the poles. The construction of a small four pole d.c. generator is shown diagramatically in Fig. 7.17; details of the bearings, brushgear, etc. have been omitted from these diagrams in order to show more clearly the main parts of the machine.

7.82 The armatures of modern machines are usually fitted with a winding consisting of coils fitted into the armature slots. In order to develop the maximum e.m.f. each coil is so shaped that its sides are spaced the same distance apart as two adjacent poles, as shown in Fig. 7.18(a). It should be noted that the coils are normally fitted with one side at the

141

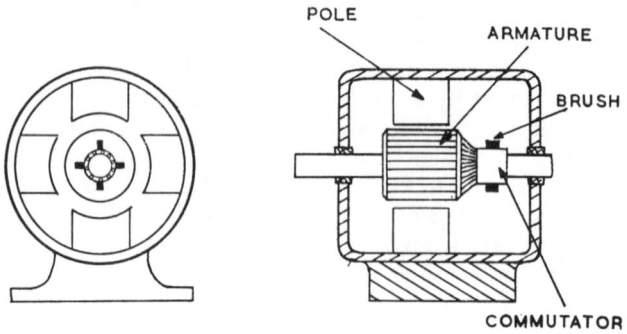

FIGURE 7.17   Four Pole d.c. Generator.

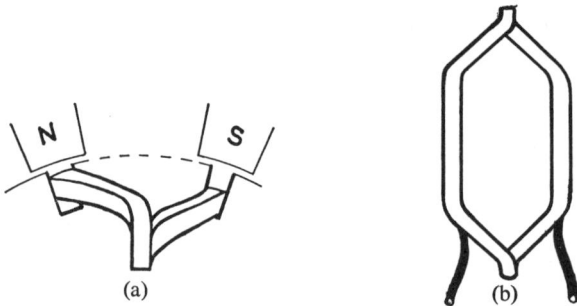

FIGURE 7.18   Armature Coils.
(a) Position of Coil.
(b) Shape of Coil.

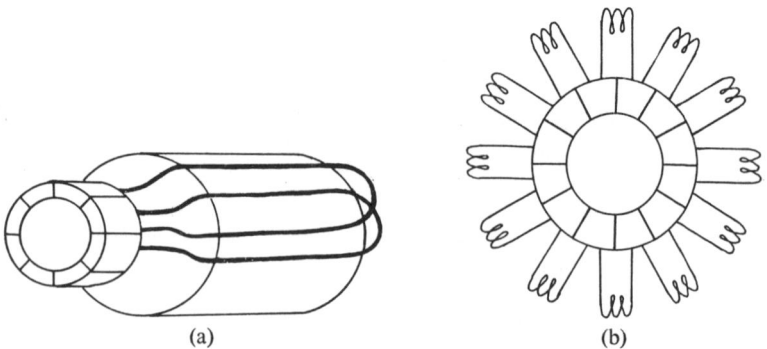

FIGURE 7.19   Lap Winding.
(a) Arrangement of Coils.
(b) Electrical Connections.

142

inside of a slot and the other at the outside, so that, when all coils are in position, the winding is completely symmetrical. Figure 7.18(b) shows the shape of a typical armature coil.

7.73 The ends from the coils are joined to the commutator segments, there being as many commutator segments as there are coils in the winding. One form of connection is known as a *lap winding* When this type of winding is employed the two connecting leads from each coil are connected to adjacent commutator segments. Figure 7.19(a) shows how the coils are arranged, while Fig. 7.19(b) shows a simplified diagram of the electrical connections. While a detailed discussion of the action of armature windings is beyond the scope of this book it can be shown that this

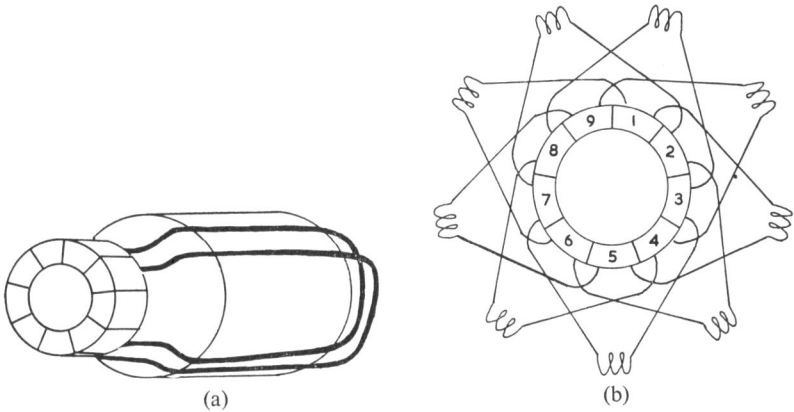

(a)   (b)

FIGURE 7.20   Wave Winding.
(a) Arrangement of Coils.
(b) Electrical Connections.

type of winding tends to provide a large current at a comparatively low voltage and is normally more suitable for the larger type of machine. An alternative method of connecting the armature coils is known as a *wave winding*. When this type of winding is employed the leads from each coil are connected to commutator segments which are approximately the same distance apart as two poles of like polarity. The wave winding, which is illustrated in Fig. 7.20 provides the maximum possible output voltage, but as it gives a lower current rating than the lap winding it is usually employed in the smaller machines.

7.84 Figure 7.21 shows the magnetic circuit of a four pole generator. The magnetomotive force required to magnetise the circuit is produced by the current flowing in the field coils mounted on the poles. The magnetic flux traverses a path through two poles, the yoke joining the poles, the air gaps between poles and armature and the armature iron itself. In order to obtain the highest possible flux density the iron parts of the circuit are

143

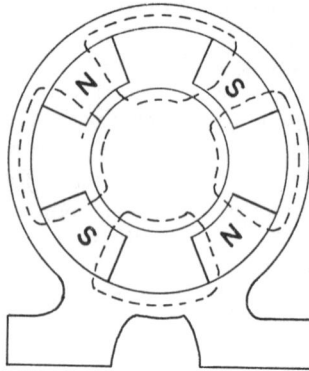

FIGURE 7.21   Magnetic Circuit of 4-Pole d.c. Generator.

constructed using material of high permeability and the air gaps are made
as short as possible. The poles themselves may be of solid material such
as mild steel or built up of silicon steel laminations; if the poles are solid
they are usually fitted with laminated pole shoes to reduce eddy current
losses in the pole faces. The yoke or magnet frame which completes the
magnetic circuit between the outer ends of the poles often consists simply
of the cast or rolled steel frame of the machine, but in some cases this is
reinforced by silicon steel laminations. The armature is built up from
silicon steel laminations, a laminated construction being necessary to
reduce eddy current losses.

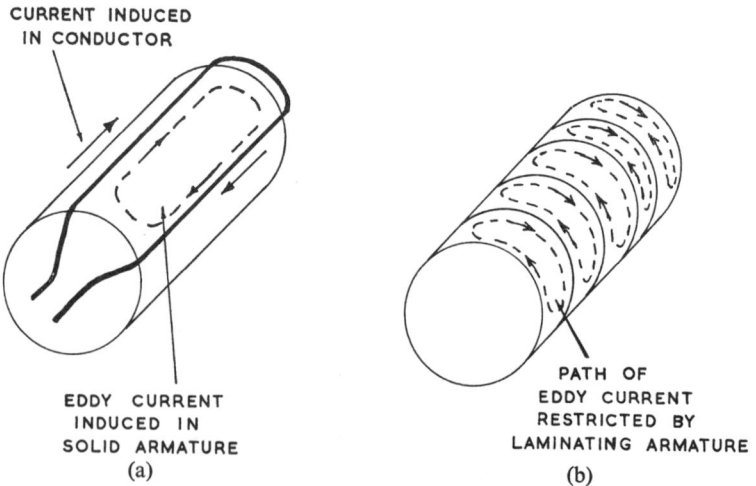

CURRENT INDUCED
IN CONDUCTOR

EDDY CURRENT
INDUCED IN
SOLID ARMATURE
(a)

PATH OF
EDDY CURRENT
RESTRICTED BY
LAMINATING ARMATURE
(b)

FIGURE 7.22   Eddy Currents in Armature.
(a) Solid Armature Provides Easy Path for Eddy Currents.
(b) Eddy Currents are Restricted by Laminated Armature.

7.85 *Eddy currents* are undesired electric currents induced in the metalwork of the machine. They heat the metalwork in which they flow and so waste power and reduce the efficiency of the machine. In the case of an armature, as well as an e.m.f. being induced in the armature conductors there will also be an e.m.f. induced in the iron of the armature as illustrated in Fig. 7.22. If the armature were solid a heavy circulating current could flow, heating the armature and wasting power. Constructing the armature of laminations prevents the easy passage of a current along the length of the armature and so considerably reduces the eddy currents. The problem of eddy currents induced by transformer action is discussed in Section 10.2.

7.9 Generator Characteristics

7.91 The characteristics of a generator depend largely on the method used to supply current to the field coils, and so the various types of generator are named in accordance with the way in which the field windings are connected. A generator in which the field windings are connected in series with the armature is called a *series* generator, while a generator in which the field windings are connected in parallel with the armature is called a *shunt* generator. A *compound* generator has two sets of field coils, one set connected in series and the other in parallel with the armature. A separately excited generator has its field coils energised by a separate d.c. supply. Figure 7.23 shows the electrical connections for the various types of generator.

7.92 It has been stated (section 7.51) that the e.m.f. induced in a conductor is proportional to the rate at which it cuts through magnetic flux. It follows that the e.m.f. developed by a generator is directly proportional to the speed with which the armature rotates and to the amount of magnetic flux produced by the poles. The flux per pole is approximately proportional to the current in the field winding. The actual output voltage available, or terminal voltage, is less than the generated e.m.f. This is because voltage drops occur owing to the effect of the resistance of the armature and any series connected field coils, and there is also a constant voltage drop of about 2 V called the brush contact drop which takes place at the brushes. The terminal voltage can be calculated by using the formula:

$$V = E - I_a R_a - I_s R_s - V_b$$

where V = terminal voltage
E = generated e.m.f.
I_a = armature current
R_a = armature resistance
I_s = current in series field coil (often the same as the armature current)
R_s = resistance of series field coil
V_b = brush contact drop

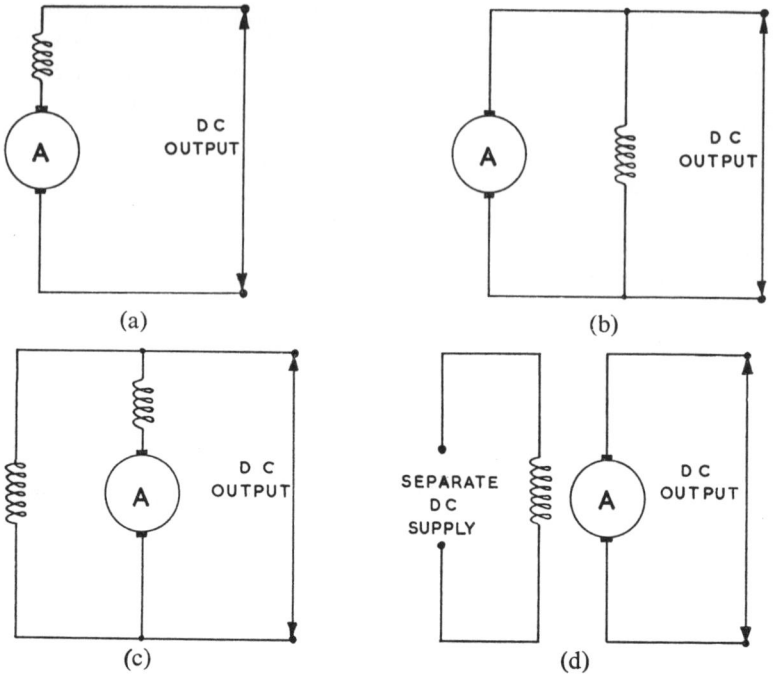

FIGURE 7.23   Connections of d.c. Generator Field Windings.
(a) Series.
(b) Shunt.
(c) Compound.
(d) Separately Excited.

Example 7.5

A d.c. shunt generator delivers a current of 96 A at 240 V. The armature resistance is 0·15 ohm, and the field winding has a resistance of 60 ohms. Assuming a brush contact drop of 2 V, calculate:
  (a) the current in the armature,
  (b) the generated e.m.f.   C.G.L.I. (part question)

Solution

(a) The connection diagram for a shunt generator, Fig. 7.24, shows that the armature must supply both the load and the shunt field currents. The shunt field current is calculated using ohm's law;

$$I_f = \frac{V}{R_f}$$

$$I_f = \frac{240}{60}$$

$$I_f = 4\,\text{A}$$

146

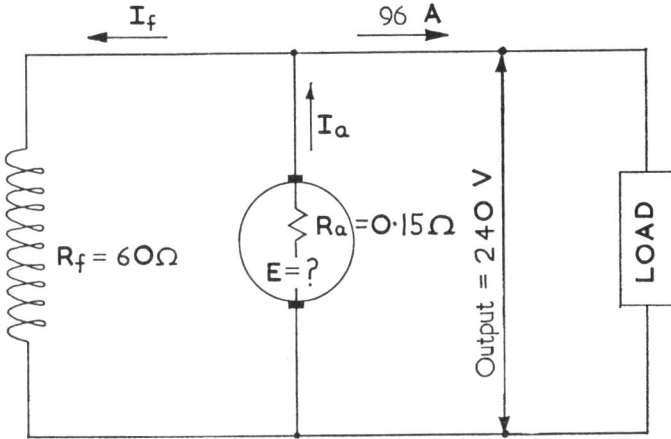

FIGURE 7.24   Example 7.5.

Hence the armature current is:

$$I_a = I_f + \text{Load current}$$
$$I_a = 4 + 96$$
$$\underline{I_a = 100 \text{ A}}$$

(b) Now $V = E - I_a R_a - V_b$. (As in this case there is no series field, the term $I_s R_s$ is not required in the equation.)

$$\therefore \quad E = V + I_a R_a + V_b$$
$$E = 240 + (100 \times 0.15) + 2$$
$$\underline{E = 257 \text{ V}}$$

7.93 In a series generator the only current flowing in the field windings is the load current so that, when the current supplied by the generator is small, the current flowing in the field coils is also small, which results in a weak magnetic field and a small generated e.m.f. At a higher value of load current the magnetic field is strengthened, giving a higher e.m.f. Thus the e.m.f. increases as the current output increases, in the manner illustrated by Fig. 7.25. The characteristics of this type of generator limit its use to such applications as a "booster" to offset the voltage drop in a long cable.

7.94 In a shunt generator the field winding is connected in parallel with the armature and so the field current is determined by the terminal voltage, and not by the load current taken from the generator. The terminal voltage of this type of generator tends to be fairly constant but drops slightly as the load current increases. This voltage drop occurs in the first instance because of the resistance of the armature windings but, as any fall in terminal voltage must reduce the field current, which in turn reduces

147

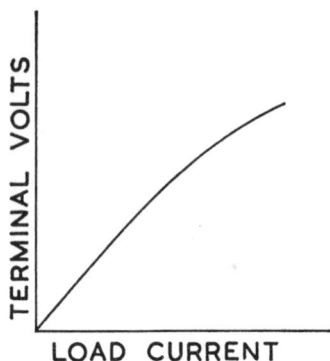

FIGURE 7.25   Series d.c. Generator Characteristic.

the generated voltage, the overall voltage drop is greater than that caused by the armature resistance alone. The characteristics of this type of generator make it suitable for the majority of applications where a fairly steady voltage is required such as small lighting plants, and battery-chargers. The terminal voltage can be easily controlled by connecting a rheostat (field regulator) in series with the field winding. When the field current is reduced (by increasing the rheostat's resistance) the generated voltage is also reduced.

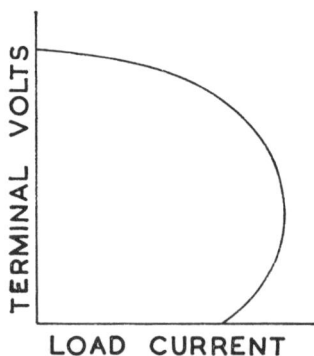

FIGURE 7.26   Shunt d.c. Generator Characteristic.

7.85   A compound generator has both shunt and series field windings. The shunt field winding usually provides most of the magnetization of the poles; the series winding can be connected to either increase or decrease the pole strength as the load current increases. When the series winding is so connected as to increase the pole strength as the load current rises the machine is known as a *cumulative compounded generator*. The effect of cumulative compounding is to increase the generated voltage as the load rises so tending to compensate for the voltage drop in the armature

148

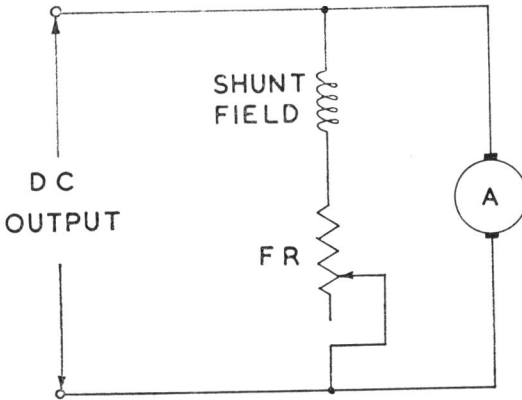

FIGURE 7.27 Field Regulator Connections.

resistance. A machine with an almost constant terminal voltage is known as a "level" compounded type. A machine giving an increase of terminal voltage with load current is known as an "over compounded" type. The characteristics of the level compounded type of generator make it suitable for applications where a more constant voltage than that provided by a shunt generator is needed. Over-compounded types are useful where a load is to be supplied via a long cable, as the rise in voltage at the generator terminals helps to offset the voltage drop in the cable as the load current

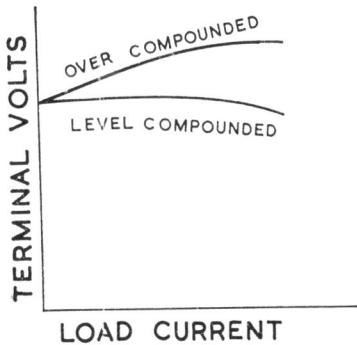

FIGURE 7.28 Cumulative Compound d.c. Generator Characteristics.

increases. When the series winding is so connected as to reduce the pole strength as the load current rises the machine is known as a *differentially compounded generator*, the effect of differential compounding being to cause the generated voltage to fall as the load current increases. Differential compounded machines are used only for special purposes such as d.c. arc welding.

7.96 The output voltage of the separately excited generator is determined mainly by the voltage of the separate d.c. supply to the field, and so can be varied smoothly over a wide range; the polarity of the generated voltage can be reversed by reversing the field current. The terminal voltage tends to fall slightly on load due to the effect of armature resistance.

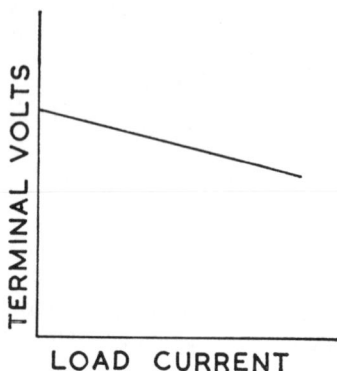

FIGURE 7.29   Separately Excited d.c. Generator Characteristic.

7.97 All self-excited generators, whether series, shunt or compound, rely upon residual magnetism to provide their initial excitation. For example, in the case of a shunt generator, when the machine is first started, the residual magnetism of the poles causes a small e.m.f. to be induced in the armature. This in turn causes current to flow in the field winding which strengthens the poles so increasing the e.m.f. The effect is progressive until the machine attains full voltage. It should be noted that the machine can only excite if:

(a) There is some initial residual magnetism.
(b) The field winding is connected so that it tends to strengthen the initial magnetism in the poles.

If a generator has lost its residual magnetism and so fails to excite, the leads from a battery may be momentarily touched on the generator terminals to provide enough magnetism to initiate excitation. The polarity of a self-excited generator depends on the polarity of the residual magnetism; it cannot be reversed merely by reversing the field connections as this would simply result in the generator failing to excite.

7.10   D.C. Motors

7.101 In paragraph 7.3 it was shown that a current-carrying coil situated in a magnetic field experiences a torque; this effect is utilised in the d.c. motor. In order to obtain continuous rotation it is necessary to reverse the direction of the current through the coil at the instant that the coil sides pass from pole to pole. This is achieved by a commutator as shown in Fig. 7.30. A practical motor uses many armature coils, and a corresponding number of commutator segments, to obtain a more even and continuous torque. The construction of a d.c. motor is similar to the

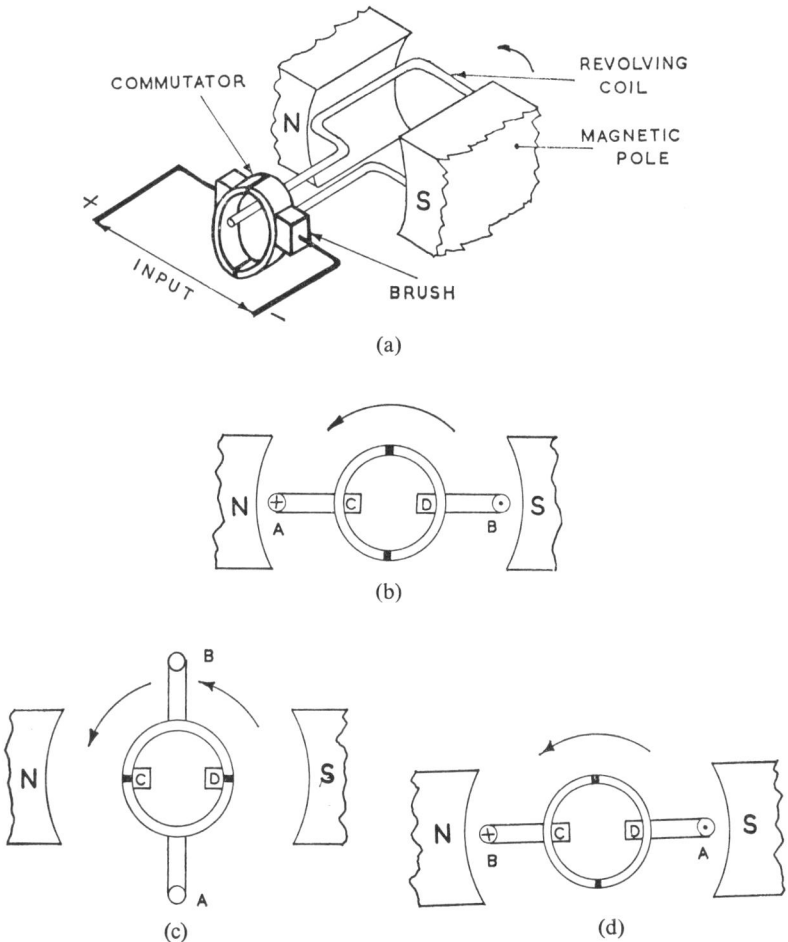

(a)

(b)

(c)   (d)

FIGURE 7.30   Simple Motor.
(a) Single Loop Coil and Commutator.
(b) Direction of Torque when Coil Side A is Opposite a North Pole.
(c) Coil Sides Between Poles, No Torque Produced.
(d) Coil Side A Opposite a South Pole, but Torque is in the Same Direction as Before.

151

construction of a generator; indeed a generator can be used as a motor and vice-versa.

7.102 When a motor armature is rotating there is an e.m.f. induced in it just as if it were a dynamo. This e.m.f. opposes the flow of current in the armature, in accordance with Lenz's law (paragraph 7.43), and is therefore called the *back e.m.f.* of the motor. It is interesting to note that when a generator is supplying a load, the armature currents produce a torque which tends to slow down the generator. The electric pressure available to overcome the armature resistance of a motor is the difference between the supply voltage and the back e.m.f., hence the armature current is given by:

$$I_a = \frac{V - E_b}{R_a}$$

where I_a = armature current
V = supply voltage
E_b = back e.m.f.
R_a = armature resistance
The equation given above can be transposed to give:

$$V = E_b + I_a R_a$$

This form of the equation shows clearly that as the armature resistance, R_a, of practical machines is normally small, the back e.m.f. nearly equals the supply voltage. As in the case of the e.m.f. developed by a generator, the back e.m.f. of a motor is directly proportional to the armature speed and to the strength of the magnetic field set up by the poles.

Example 7.6

Explain the meaning of *back e.m.f.* of a d.c. motor and explain how it is produced.

A 460 V d.c. shunt motor, running on load, has an armature resistance of 0·12 ohm. Calculate:

(i) the value of the back e.m.f. when the current in the armature is 150 A,
(ii) the value of the armature current when the back e.m.f. is 452 V.

C.G.L.I.

Solution

The back e.m.f. of a d.c. motor is an e.m.f. induced in the armature windings which opposes the flow of current through the windings. This e.m.f. is produced by the action of the armature conductors cutting through the magnetic flux produced by the poles, in the same way that an e.m.f. is produced in a generator.

152

(i)
$$V = E_b + I_a R_a$$
$$\therefore \quad E_b = V - I_a R_a$$
$$E_b = 460 - 150 \times 0\cdot12$$
$$E_b = 460 - 18$$
$$\underline{E_b = 442 \text{ V}}$$

(ii)
$$I_a = \frac{V - E_b}{R_a}$$
$$I_a = \frac{460 - 452}{0\cdot12}$$
$$\underline{I_a = 66\cdot7 \text{ A}}$$

7.103 The torque produced by a motor is proportional to the armature current and to the magnetic flux per pole. If the torque produced by a motor is more than that required to drive the load it will accelerate, so increasing the back e.m.f. and reducing the armature current until, eventually, just enough current is taken to produce the load torque. If, on the other hand, a motor produces less torque than that required by the load it will slow down, thus reducing the back e.m.f. and allowing more current to flow through the armature until once again the torque produced by the armature current balances the load torque. The current consumed by any motor will be that required to produce the load torque, while the speed will automatically adjust itself to the value required to give the correct back e.m.f. If the magnetic field of a motor is weakened the motor must run faster to maintain its back e.m.f., and if the magnetic field is strengthened the motor will slow down.

7.104 In a series motor the armature current also passes through the field winding so the magnetic field is strong when the motor is heavily loaded, giving rise to a high torque at a low speed. If the load is reduced

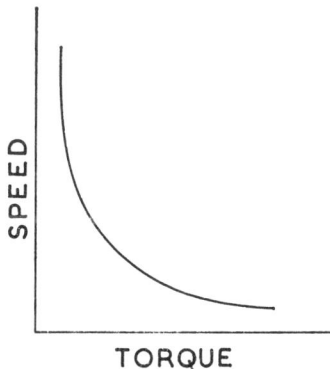

FIGURE 7.31   Series Motor Characteristic.

the magnetic field is weakened so the speed increases; at very light loads the speed can increase to a dangerously high value. The characteristics of this motor make it suitable for applications where a high starting torque is required, and where the speed variation with load is no disadvantage. The motor must always be directly coupled to the load as the speed would become excessive if the load were suddenly removed. This makes the motor ideal for traction purposes, a further advantage in this case being that the high starting torque is still available even if there is considerable voltage drop in the supply mains.

7.105 In shunt motors the field winding receives the full supply voltage and so the strength of the magnetic field is constant, hence the speed is fairly constant, falling slightly at high loads, owing to the effect of armature resistance. The speed can be easily controlled by using a field rheostat

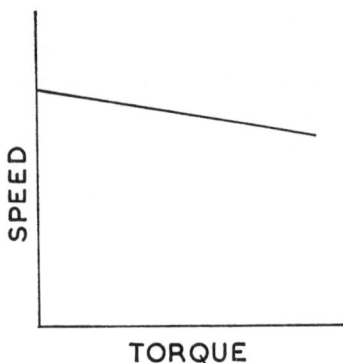

FIGURE 7.32   Shunt Motor Characteristic.

as explained in paragraph 7.112 below. The characteristics of this motor make it suitable for situations where a fairly constant speed is required such as driving machine-tools, line shafting, etc. It is also useful where a moderate degree of speed control is required.

7.106 The compound motor has both shunt and series field windings and, like the generator, either cumulative or differential compounding may be employed. If the effect of the series winding is to increase the pole strength as the load current increases this is known as cumulative compounding, and the motor characteristics are intermediate between the series and shunt types, giving a high starting torque together with a safe no-load speed. These factors make the cumulative compound motor suitable for driving heavy intermittent loads such as lifts, hoists, fly presses, etc. If the effect of the series winding is to weaken the pole strength as the load current increases this is known as differential compounding. The effect of differential compounding is that, as the load increases the poles are weakened, so tending to raise the motor speed and thus compensating for the natural speed drop due to the increase in load.

154

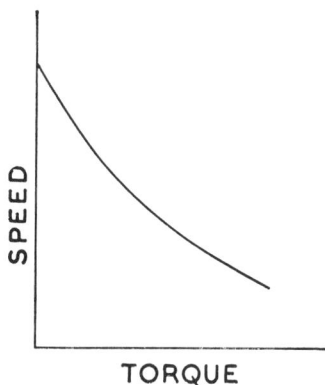

FIGURE 7.33   Cumulative Compound Motor Characteristics.

Thus a differential compounded motor can have a constant speed characteristic. Such motors are seldom used, as they have a tendency to become
unstable. A slight excess of the differential compound effect can cause the
motor speed to rise uncontrollably up to a dangerously high level.

7.11   Speed Control

7.111   In general there are two methods of controlling the speed of
d.c. motors; these are:
(a) By varying the voltage applied to the armature.
(b) By varying the strength of the magnetic field.
A rheostat, connected in series with the armature, will reduce the voltage
applied to the armature, thereby lowering the speed. Unfortunately this
method of speed control has the following serious disadvantages:
(a) The rheostat must be capable of carrying the full armature current,
and so tends to be expensive.
(b) The power loss in the rheostat lowers the overall efficiency.
(c) The effectiveness of the rheostat varies with the load current.
7.112   A widely used method of controlling the speed of shunt and
compound motors is to connect a rheostat in series with the shunt field as
shown in Fig. 7.34. When the rheostat is adjusted so as to increase its
resistance the field current is decreased, so weakening the magnetic
field. The motor now has to run faster to maintain the back e.m.f. It is
necessary to select the motor so that its speed, with the rheostat at its
minimum setting, is the lowest that may be required, as an increase in
rheostat resistance only increases the speed. A speed variation of approximately 3 to 1 can be obtained with average machines without weakening
the poles unduly; a wider speed range usually calls for a specially constructed machine.

155

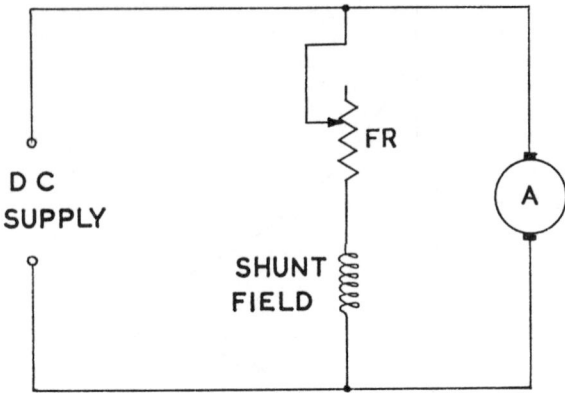

FIGURE 7.34. Speed Control of Shunt Motor.

7.113 The speed of a series motor may be regulated by using divertor resistors connected as shown in Fig. 7.35. The field divertor reduces the current in the series field so increasing the speed of the motor, while the armature divertor increases the current in the series field thereby reducing the speed of the motor.

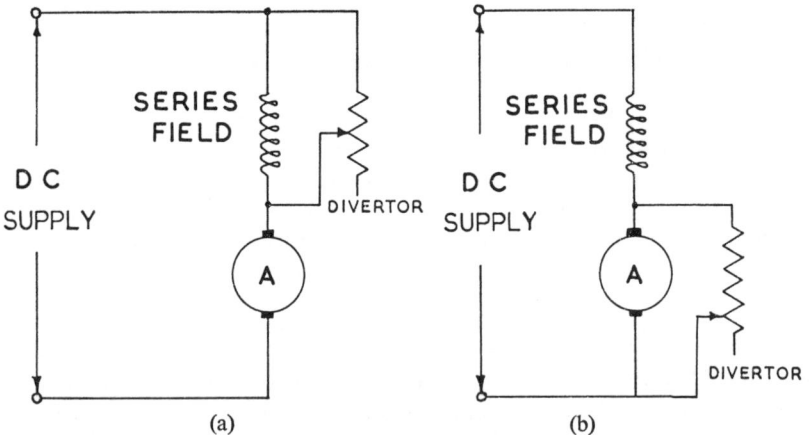

FIGURE 7.35 Speed Control of Series Motor.
(a) Field Divertor.
(b) Armature Divertor.

7.12 Reversal of D.C. Motors

7.121 It is sometimes necessary to reverse the direction of rotation of a d.c. motor. By considering the left hand rule (section 7.2) it can be seen that the direction of the force acting on a conductor is reversed if either

156

the direction of the current or the magnetic flux is reversed, but not if both are reversed. It follows that in order to reverse the direction of rotation of any d.c. motor it is necessary to reverse the connections of either the field winding or the armature but not both. Figure 7.36 shows the alterations in connections required for reversed rotation of series, shunt and compound motors.

MOTOR TYPE	ORIGINAL CONNECTION	FIELD REVERSED	ARMATURE REVERSED
SERIES			
SHUNT			
COMPOUND			

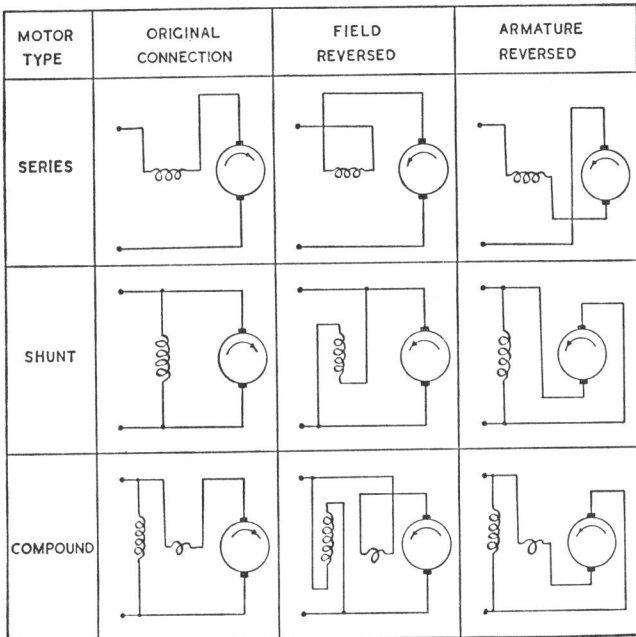

FIGURE 7.36   Reversal of d.c. motors.

7.13   Motor Starting

7.131   When a motor is first switched on there is no back e.m.f. to limit the armature current so that, if no method of limiting the starting current is employed, this current often becomes large enough to cause damage to the motor and its associated wiring. Also, excessive voltage drops may be caused in the supply mains thereby affecting other apparatus. For example, if a 500 V d.c. motor having an armature resistance of 0·02 ohms is connected directly to the supply then the initial starting current will be:

$$I = \frac{500}{0·02} = 25\ 000\ \text{A}$$

A good method of reducing the starting current is to connect a rheostat in series with the armature, the rheostat being gradually cut out of circuit as the motor gathers speed so developing a back e.m.f.

157

7.132 Figure 7.37(a) shows a simplified diagram of a d.c. motor starter. When the starter arm is moved to the first stud the current taken by the motor immediately jumps to a value limited by the starter resistance plus armature resistance. The motor then starts to rotate and, as it accelerates, the back e.m.f. increases so reducing the current. After a short time the starter arm is moved to the next stud; the current immediately increases as some resistance has been cut out of the circuit and then gradually falls as the motor speed increases yet again. Starting resistors are cut out of circuit one by one until the motor is up to full speed and is developing its full back e.m.f. with no added resistance in circuit. Figure 7.37(b) shows graphically the manner in which the motor current varies during the starting operation.

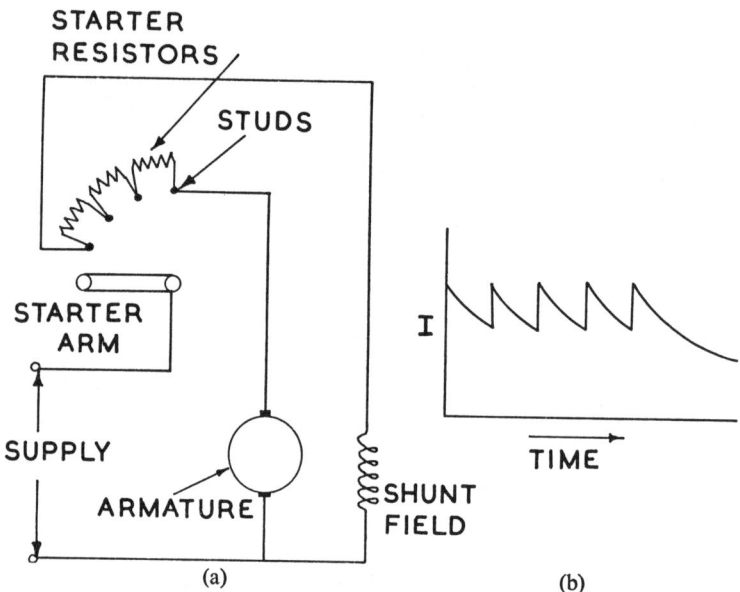

FIGURE 7.37   Principle of Motor Starter.
(a) Simplified Diagram of Starting Circuit.
(b) Variation of Starting Current.

7.133 In addition to limiting the starting current, motor starters should include safety devices to:

(a) Automatically re-connect the starting resistors whenever the motor is stopped.

(b) Prevent automatic re-starting should the motor stop owing to supply failure.

(c) Automatically disconnect the motor in the event of a sustained overload.

158

A good example of simple motor starter incorporating devices to fulfil the above requirements is the face-plate starter illustrated in Fig. 7.38. The starter arm (S.A.) is spring loaded to return to the OFF position, and when in this position the supply to the motor is interrupted. When the starter arm is moved sufficiently to make contact with the first stud the field winding receives the full supply voltage but the armature current is limited by the starting resistors. As the motor gathers speed the starter arm is gradually moved to the ON position, so cutting out the starting resistors from the armature circuit. The starter arm is held in the ON

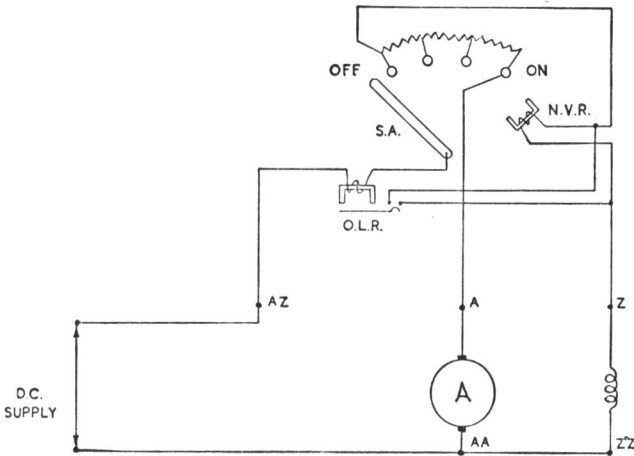

FIGURE 7.38 Face Plate Starter.

position by the magnetic attraction of the *no-volt release* (N.V.R.) which is energised by the field current. Should the supply voltage fail, or the field circuit be broken, the no-volt release is de-energised and the starter arm flies back to the OFF position under the influence of its spring. Thus the no-volt release provides a means of preventing automatic re-starting after a mains failure, ensures that the starter arm returns to OFF when the motor is switched off and also protects against the consequences of an open circuit occurring in the field winding while the motor is running. The current consumed by the motor flows through the *overload relay* (O.L.R.) and if this current becomes excessive, for example due to overloading the motor, the relay armature is attracted. Movement of the relay armature closes contacts which short circuit the no-volt release. This again allows the starter arm to fly back to the OFF position, so switching off the motor.

SUMMARY OF FORMULAE

Force on conductor: $F = BIl$
Torque produced by coil: $T = BINA$

159

Induced e.m.fs.:
$$e = \frac{\text{Change in magnetic flux linkage}}{\text{Time taken to change}}$$

$$e = Blv$$

$$e = -L \times \text{rate of change of current.}$$

E.M.F. induced in coil $A = M \times$ rate of change of current in coil B.

Energy stored in inductor:
$$W = \tfrac{1}{2}LI^2$$

Time constant of LR combination:
$$\tau = \frac{L}{R}$$

Terminal voltage of d.c. generator:
$$V = E - I_a R_a - I_s R_s - V_b$$

Motor armature current:
$$I_a = \frac{V - E_b}{R_a}$$

EXERCISE 7

1. Sketch a conductor moving through a magnetic field at right angles to the direction of flux. Show the directions of motion, flux and induced e.m.f.

   A conductor of length 200 mm moves at 80 metres/second at right angles to a magnetic flux of density 1·2 Wb/m² (or 1·2 tesla). Calculate the e.m.f. induced in the conductor. Explain how the e.m.f. may be reduced by one half.

   E.M.E.U.

2. (a) Describe, with the aid of sketches, how a force is exerted on a current carrying conductor when placed in, and at right angles to, a magnetic field.
   (b) Calculate the force exerted on a conductor 3 metres long when carrying a current of 10 A and situated in, and at right angles to, a magnetic field of flux density 0·8 Wb/m².

   N.C.T.E.C.

3. (a) Consider a single conductor carrying direct current and lying in the magnetic field between the poles of a 2-pole d.c. motor. Make a neat sketch or diagram illustrating the above. Assuming your own directions of magnetic field and current, indicate clearly the direction in which the conductor will tend to move.
   (b) A conductor 300 mm long lies at right angles to a magnetic field of intensity 1·6 webers per metre², (or 1·6 tesla), and carries a current of 25 A. Calculate the force on the conductor.

   C.G.L.I.

4. A conductor of length 80 mm moves at a speed of 2000 mm/s at right angles to a uniform magnetic field of flux density 0·6 Wb/m², (or 0·6 T). Determine the e.m.f. induced in the conductor.

   Explain how the e.m.f. may be increased.

   Draw a sketch to show the directions of the motion, e.m.f. and magnetic field.

   E.M.E.U.

5. With the aid of sketches and diagrams, describe and explain how an alternating current may be generated in a simple 2-pole generator. Assume that the armature consists of a single coil of wire.

   What are the effects of an alteration of (a) the driving speed, (b) the excitation?

   C.G.L.I.

6. Describe, with suitable sketches, a 4-pole direct-current shunt-wound generator, naming the different parts and explaining the function of each.

   Explain briefly how the d.c. generator works.

   C.G.L.I.

7. A d.c. shunt excited generator, having a field resistance of 240 ohm and an armature resistance of 0.5 ohm, supplies a load of 6 kW at 240 V. Calculate the e.m.f. generated in the armature.
    U.L.C.I.

8. Draw simple sketches illustrating the method of connecting the field coils to the armature in the case of (a) series, (b) shunt and (c) compound wound machines.
    What effect does increasing the load have on the terminal voltage of each of these machines? Illustrate your answer with characteristic curves.   N.C.T.E.C.

9. (a) Explain briefly, with the aid of diagrams, the differences between series, shunt and compound d.c. generators.
    (b) A shunt-wound d.c. machine has armature resistance of 0·2 ohm, and a shunt resistance of 115 ohms. It is used (i) as a shunt generator to deliver a current of 150 A to an outside load, and (ii) as a shunt motor driving a mechanical load and taking 150 A from a d.c. supply. In both cases the terminal voltage of the machine is 460 V.
    Calculate the currents in both armature and field windings in both cases (i) and (ii).
    C.G.L.I.

10. Describe the construction of a compound wound shunt generator. What is the object in using a compound wound generator? If the machine is run at a constant speed, how can the voltage at the armature be regulated?
    Sketch a graph of variation of voltage with load current when the machine is (a) not compounded, (b) compounded.
    If the current in the armature and series field winding is 20 A and the combined resistance is 0·75 Ω, calculate the power lost in these windings.   U.L.C.I.

11. What is the meaning of the expression *back e.m.f.* of a direct-current motor?
    Explain how the back e.m.f. and the current change during the starting of a d.c. motor.
    C.G.L.I.

12. Explain why the armature current of a d.c. motor rises when an increased load is placed on the shaft.
    A 250-V d.c. motor runs on no load with armature current 20 A; the armature resistance is 0·5 ohm. Determine the armature generated (back) e.m.f.
    If on full-load, the speed falls 10%, determine the new generated (back) e.m.f. and armature current with flux per pole constant.   E.M.E.U.

8 Alternating Currents I

8.1 Alternating Quantities

8.11 An alternating quantity is one which periodically reverses its direction of action. The quantity concerned goes through a sequence of events called a *cycle* in a definite time, and then repeats the same sequence in the same time over and over again. Figure 8.1 shows a graph of the values of an alternating quantity plotted against time. The shape of this graph is called the waveform of the quantity.

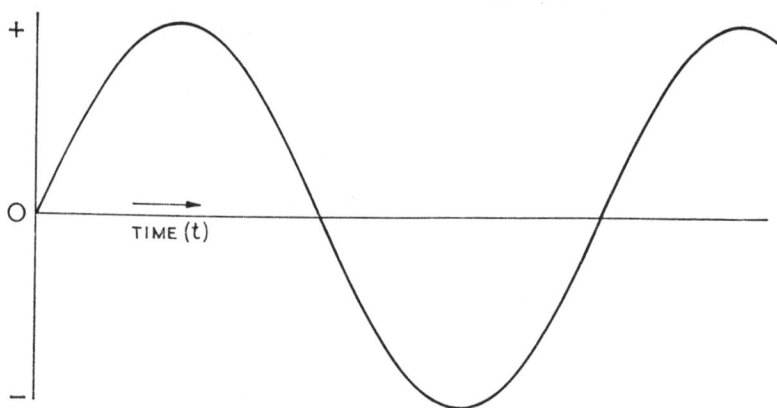

FIGURE 8.1   Waveform of an Alternating Quantity.

8.12 The time taken to complete one cycle is called the *periodic time* (symbol T). The number of cycles completed in a time of one second is called the *frequency* (symbol f), the unit in which this is measured is called the **hertz** (abbreviation Hz).* The relation between periodic time and frequency is given by the formula:

$$f = \frac{1}{T}$$
$$\text{or} \quad T = \frac{1}{f}$$

* In the past the symbols c.p.s. and c/s have been widely used instead of Hz.

162

8.2 Measurement of Alternating Quantities

8.21 The measurement of an alternating quantity presents some difficulties as the quantity concerned is continually changing both its magnitude and direction. Because of this there are three important methods in common use for denoting the magnitude of a waveform by means of a single figure or constant.

8.22 The *peak* or *maximum* value of a waveform is the value of the highest point of the waveform.

8.23 The *average* or *mean* value of a waveform is the average value of the waveform reckoned over one half of a complete cycle. If an attempt is made to average a purely alternating waveform over a complete cycle, the negative half of the wave cancels the positive half exactly, and the result is zero.

8.24 The *root-mean-square* (r.m.s.) value of an electric current or pressure is a measure of how useful it is in practice. As the heating effect of an electric current operates equally well for a.c. or d.c. supplies, the r.m.s. value is such that a d.c. current or pressure of this value would produce the same heating effect as the alternating current or pressure concerned. The r.m.s. value is also sometimes called the *effective* or *virtual* value.

8.25 Because it indicates the usefulness of a supply, the r.m.s. value is the value normally used for measuring purposes; almost all a.c. electrical instruments indicate r.m.s. values. In electrical formulae involving alternating quantities symbols without suffixes are normally used to represent r.m.s. values, e.g. I, E, V. Maximum and average values are identified where necessary by the use of appropriate suffixes, e.g. I_{max}, E_{max}, V_{max}, I_{av}, E_{av}, V_{av}.

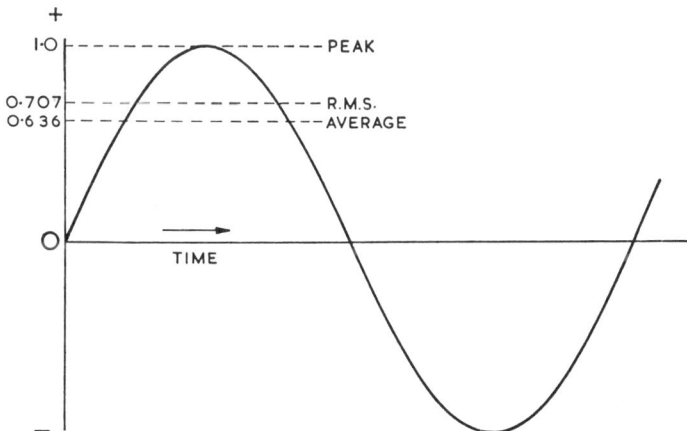

FIGURE 8.2   Values of Sine Wave.

8.3 Sine Waves

8.31 The electrical currents and voltages used for practical mains supplies have the type of waveform known as a *sine wave*. Figure 8.3 shows a graphical method of constructing a sine wave by projection from a rotating line *OA*. The line *OA* is assumed to rotate in an anticlockwise direction. The height of the sine wave at any angle is the same as the height of the tip of the line and so can be found by projecting a horizontal line across the diagram. For example when the line has turned through 30° it is in position *OA'*, and the 30° point on the sine wave is found by projection as shown in Fig. 8.3. By projecting from a number of points the complete sine wave is easily determined.

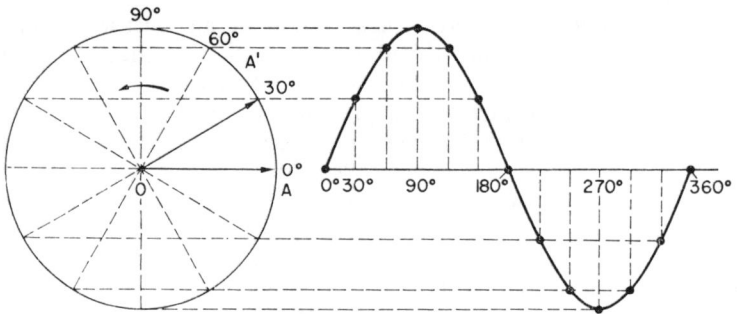

FIGURE 8.3   Construction of Sine Wave.

8.32 It can be shown mathematically that for a sine waveform

$$\text{Average Value} = \frac{2}{\pi} \times \text{Maximum Value}$$
$$= 0 \cdot 636 \times \text{Maximum Value}$$

$$\text{R.M.S. Value} = \frac{1}{\sqrt{2}} \times \text{Maximum Value}$$
$$= 0 \cdot 7071 \times \text{Maximum Value}$$

For any waveform the *form factor* is defined by the formula:

$$\text{Form factor} = \frac{\text{R.M.S. Value}}{\text{Average Value}}$$

For a sine wave the form factor $= \dfrac{0 \cdot 7071}{0 \cdot 636} = 1 \cdot 11$

8.33 Two sine waves may not be exactly in step. Figure 8.4 shows the waveforms for the e.m.f. and current in a certain circuit. The current waveform starts later than the e.m.f. waveform and so the current is said to *lag* the e.m.f. It could equally well be said that the e.m.f. *leads* the current. It is convenient to divide the cycle into 360° so that the amount of *lead* or *lag* can be expressed as a *phase-angle* (symbol ϕ).

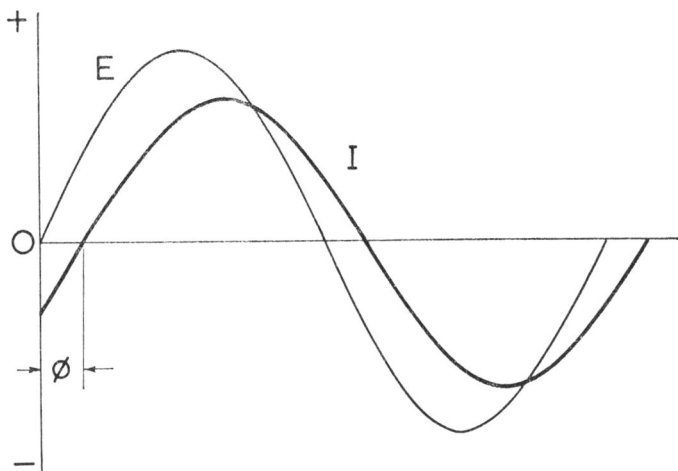

FIGURE 8.4   Phase Angle.

8.4   Phasor Representation of Sine Waves

8.41 It is often convenient to represent an alternating quantity by a phasor*, that is by drawing a straight line whose length denotes the magnitude of the quantity to scale, and whose direction represents the phase angle. A wave commencing at the origin has a zero phase angle and is denoted by a horizontal line pointing to the right. A wave commencing before the origin has a leading phase angle which is measured anticlockwise from the horizontal, while a wave commencing after the origin has a lagging phase angle which is measured clockwise from the horizontal. Figure 8.5 shows wave and phasor diagrams for these three cases.

8.42 When a set of wave diagrams for all the e.m.fs., p.ds., and currents in a circuit are drawn the origin of the graph can be made to coincide with any chosen one of the alternating quantities in the circuit. The phasor which represents this chosen quantity is drawn horizontally and is called the *reference phasor*. All phase angles must be measured from the reference phasor.

* In the past the term *vector* has often been used instead of *phasor*, and more recently the term *complexor* has been advocated.

165

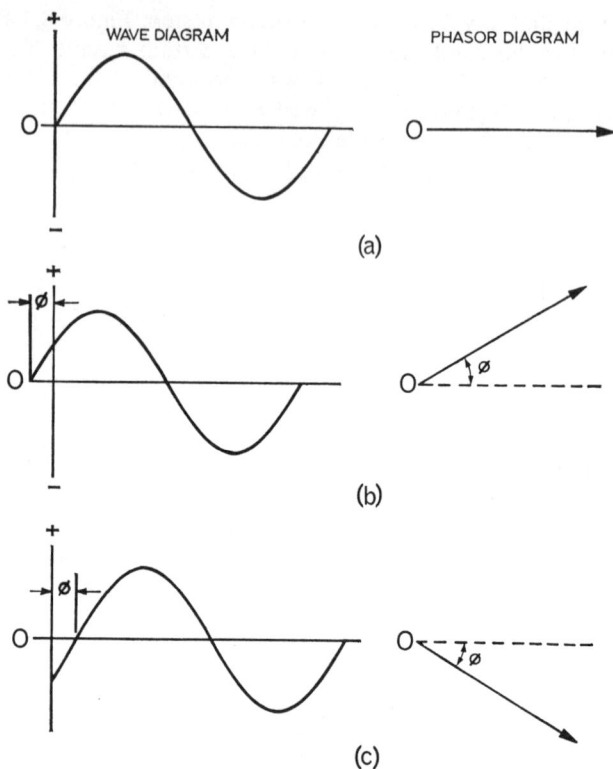

FIGURE 8.5 Phasor Representation.
(a) Zero Phase Angle.
(b) Leading Phase Angle.
(c) Lagging Phase Angle.

Example 8.1

When an alternating e.m.f. is applied to a circuit the current produced lags the e.m.f. by 60°.

Illustrate the above statement by means of wave and phasor diagrams.

Solution

There are two possible solutions both equally correct. Figure 8.6(a) gives the solution when the e.m.f. is chosen as the reference phasor, while Fig. 8.6(b) shows the result of choosing the current as the reference phasor.

8.43 When choosing the reference phasor it is usually convenient to select a quantity which has the same value in all parts of the circuit. For example, in a series circuit the same current flows in each part and the current, therefore, is taken as the reference phasor. In a parallel circuit the supply e.m.f. is the same for each branch and so, in this case, the supply e.m.f. provides a convenient reference phasor.

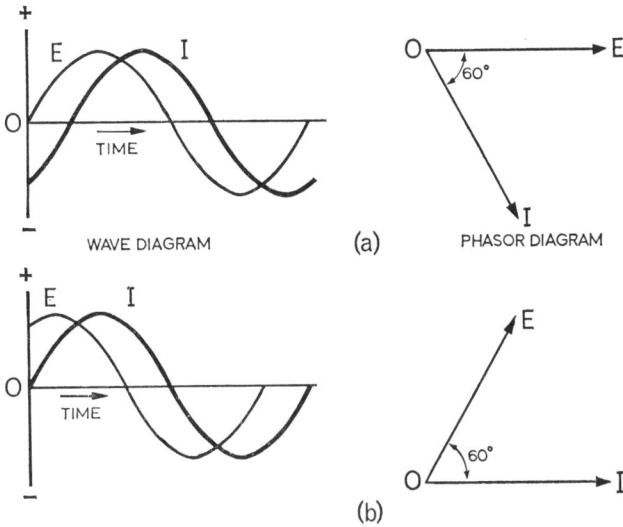

FIGURE 8.6   Example 8.1.
(a) E.m.f. as Reference Phasor.
(b) Current as Reference Phasor.

8.5   Phasor Addition

8.51   Alternating quantities cannot be added arithmetically; for example if two alternating e.m.fs. are connected in series they may not be in phase with each other. In this case the resulting e.m.f. wave could be found by adding the values at different instants of time as shown in Fig. 8.7(a).

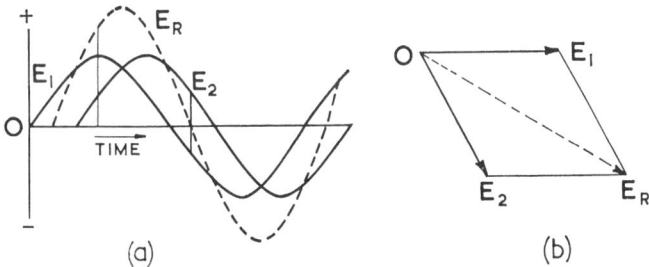

FIGURE 8.7   Addition of a.c. Waves.
(a) Using Wave Diagram.   (b) Using Phasor Diagram.

8.52   A better way of finding the resultant of two alternating quantities is to represent them by phasors as shown in Fig. 8.7(b). The resultant can then be found using the method of the parallelogram of forces already described in Section 4.6.

167

Example 8.2

Two sources of e.m.f., *A* and *B* are connected in series. Source *A* provides an e.m.f. of 100 V leading the current by 45° and source *B* provides an e.m.f. of 150 V lagging the current by 30°. Find the resultant total e.m.f.

Solution

As the two sources are connected in series they must both carry the same current; and the resultant e.m.f. is the phasor sum of the two given e.m.fs. In order to determine the phasor sum a phasor diagram can be drawn to scale. The separate steps required to build up this diagram are listed below:

Step 1. Draw the current phasor horizontally as "reference".

O————————▶I

Step 2. Draw the phasors for E_A and E_B to scale; measuring the phase angles from the reference phasor.

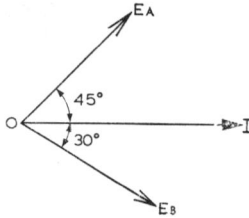

Step 3. Construct the phasor parallelogram, and hence determine the resultant total e.m.f. This is represented by the diagonal of the parallelogram of which OE_A and OE_B are the sides.

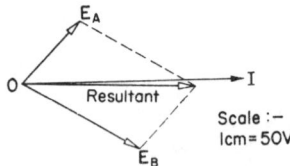

Scale :— 1cm = 50V

Answer: Resultant total e.m.f. = 201 V

8.6 Resistance in A.C. Circuits

8.61 The current flowing through a pure resistance is governed by Ohm's law for every instant of time, i.e. $i = e/R$ at every part of the cycle. This means that the current waveform for a purely resistive circuit is exactly the same shape as the waveform of the applied e.m.f., and is in

168

phase with it. Also the magnitude of the current can be found by using the expression,

$$I = \frac{E}{R},$$

where I and E are the r.m.s. values of the current and e.m.f. respectively.

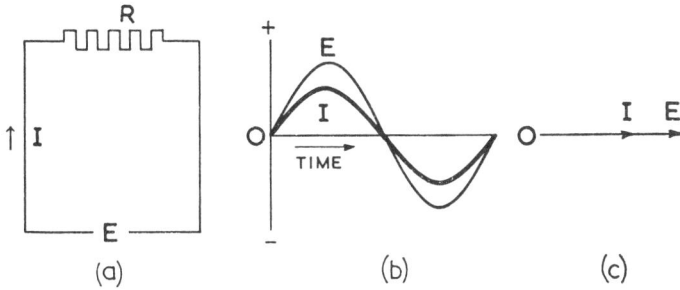

FIGURE 8.8   Purely Resistive Circuit.
(a) Circuit Diagram.
(b) Wave Diagram.
(c) Phasor Diagram.

8.62   The r.m.s. value of an alternating current is defined as the value of direct current which would produce the same heating effect when passed through a pure resistor. It follows that the power dissipated in a pure resistor is given by: $P = RI^2$ just as in a d.c. circuit.

8.7   Inductance in A.C. Circuits

8.71   When an alternating current flows through a pure inductor, the value of the current is perpetually changing so producing a self-induced e.m.f. at every instant. This self-induced e.m.f. provides the only opposition to the flow of current, and so the current will rise to a value where the induced e.m.f. is exactly equal, but opposite, to the supply e.m.f. at every instant. Figure 8.9(a) shows a purely inductive circuit supplied with an alternating e.m.f. of maximum value E_{max} and carrying a current of maximum value I_{max}. The current rises from zero to its maximum value in a quarter of a cycle i.e. in $1/4f$ seconds. It follows that the average rate of change of current is

$$\frac{I_{max}}{1/4f}$$

$\therefore$ Average rate of change of current $= 4fI_{max}$.

But the induced e.m.f., $e = -L \times$ rate of change of current, and the average supply pressure, E_{av}, is equal to the average self-induced e.m.f. but of opposite sign;

$$\therefore \quad E_{av} = 4fLI_{max}$$

169

Now, for a sine wave,

$$E_{av} = \frac{2}{\pi} \times E_{max}$$

and so

$$E_{max} = \frac{\pi}{2} \times 4fLI_{max}$$

whence

$$\frac{E_{max}}{I_{max}} = 2\pi fL$$

Converting to r.m.s. values

$$\frac{E_{max} \times 0\cdot7071}{I_{max} \times 0\cdot7071} = 2\pi fL$$

$$\frac{E}{I} = 2\pi fL$$

Now in any circuit the ratio volts/amperes is measured in ohms, so the inductor acts as if it had a value in ohms equal to $2\pi fL$. This value of ohms is called the *reactance* of the inductor (symbol X_L).

$$X_L = \frac{E}{I} = 2\pi fL$$

where: X_L = reactance of the inductor in ohms
          f = frequency in Hz
          L = inductance in henrys.

The formula shows that inductive reactance, X_L, depends not only on the value of inductance but also on the frequency of the supply. A graph of X_L plotted against frequency is a straight line as shown in Fig. 8.9(c).

8.72 The maximum induced e.m.f. in an inductor will occur at the instant when the rate of change of current is a maximum, and the polarity of this e.m.f. will oppose the change in the current. Figure 8.10(a) shows the relation between the current, induced e.m.f. and supply e.m.f. for a purely inductive circuit. The current is changing most rapidly when it passes through zero and so the maximum e.m.fs. coincide with the current zero. The supply voltage is always equal in magnitude but of opposite sign to the induced e.m.f. and, as can be seen from the wave diagram, the result is that the current in the inductor lags the supply e.m.f. by 90°. Figure 8.10(b) shows how this is represented by a phasor diagram.

8.73 The power input to a pure inductor at any instant is the product of the instantaneous values of e.m.f. and current. Figure 8.11 shows the e.m.f. and current waves for a purely inductive circuit together with the waveform of the power consumed. It can be seen that the sign of the power wave reverses every quarter of a cycle, showing that energy is being alternately fed into and taken out from the inductor. Over a complete

170

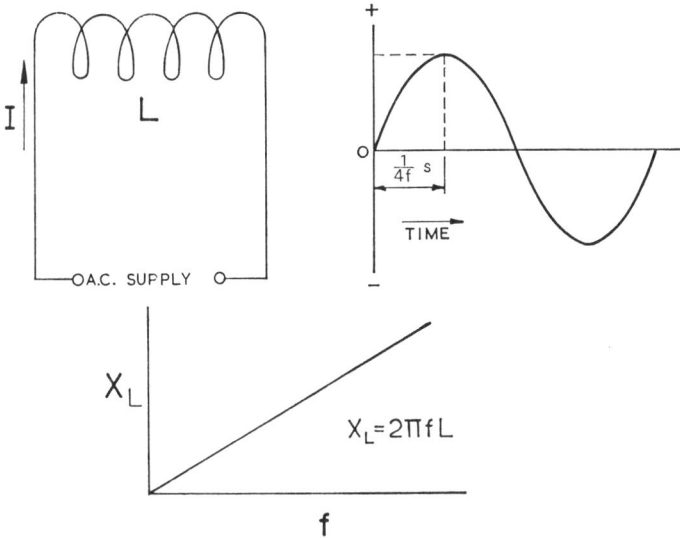

FIGURE 8.9   Inductive Circuit.
(a) Circuit Diagram.
(b) Current in Inductive Circuit.
(c) $X_L = 2\pi fL$.

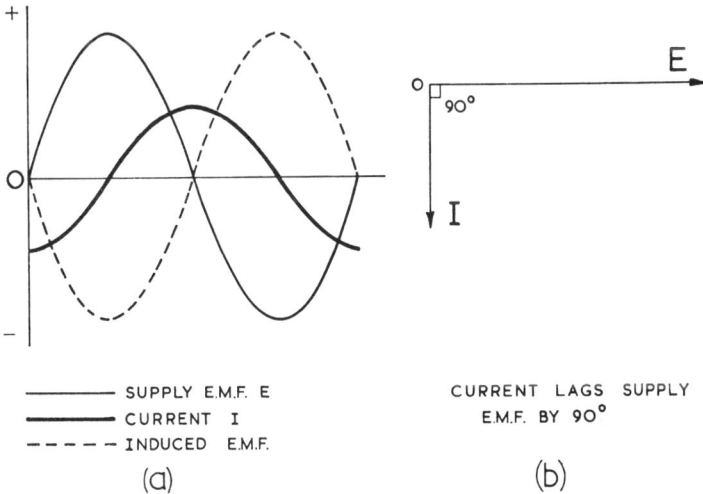

FIGURE 8.10   Current and e.m.f. in Inductive Circuit.
(a) Wave Diagram.
(b) Phasor Diagram.

171

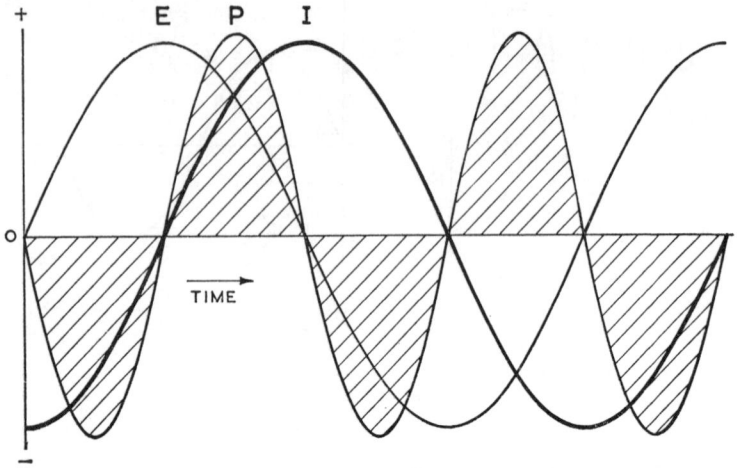

A PURE INDUCTOR CONSUMES NO POWER

FIGURE 8.11   Power Consumption of an Inductor.

cycle the positive and negative parts of the power wave are equal in area showing that the average power is zero. This gives the important result that the *average power consumed by a pure inductor is zero*.

8.8   Resistance and Inductance in Series

8.81   Figure 8.12 shows a simple series circuit in which the supply e.m.f. (E) forces a current (I) through a resistor and pure inductor. The voltmeters V_R, V_L and E register the p.ds. across R and L and the supply pressure. As these are alternating p.ds., the value of E can be determined

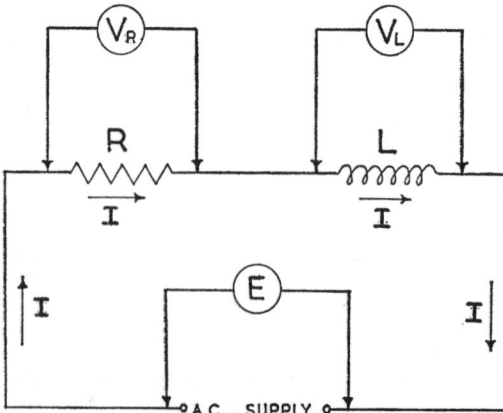

FIGURE 8.12   Simple Series Circuit.

by finding the phasor sum of V_R and V_L. The required phasor diagram is shown in Fig. 8.13 and it should be noted that:

(a) As the current is the same throughout a series circuit, I is chosen as the reference phasor.
(b) V_R is in phase with I.
(c) V_L leads I by 90° (so that I lags V_L by 90°).
(d) E is found by completing the parallelogram of phasors.

It can be seen that the triangle oab in the phasor diagram is right angled and so, by Pythagoras's theorem,

$$E^2 = V_R{}^2 + V_L{}^2$$

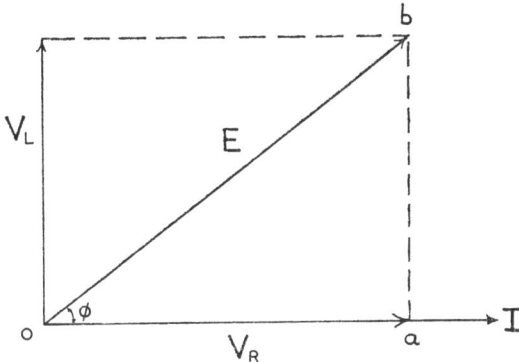

FIGURE 8.13   Phasor Diagram for Series Circuit.

8.82   As $V_R = RI$ and $V_L = X_L I$ where $X_L = 2\pi f L$, it follows that

$$E^2 = I^2(R^2 + X_L{}^2)$$

i.e.
$$\frac{E^2}{I^2} = R^2 + X_L{}^2$$

whence
$$\frac{E}{I} = \sqrt{R^2 + X_L{}^2}$$

8.83   The total opposition afforded by an electric circuit to the passage of alternating current is called its *impedance* (symbol Z). The impedance is measured in ohms and is given by:

$$Z = \frac{E}{I}$$

The expression derived in section 8.82 above, shows that for resistance and inductance in series

$$Z = \sqrt{R^2 + X_L{}^2}$$

173

Example 8.3

A single phase a.c. supply at 240 V, 50 Hz is applied to a series circuit consisting of a pure inductor of 0·05 H and a non-inductive resistor of 25 Ω. Calculate:

(a) the impedance of the circuit,
(b) the current in the circuit,
(c) the p.ds. across the inductor and the resistor and
(d) draw a phasor diagram, to scale, for the above circuit.

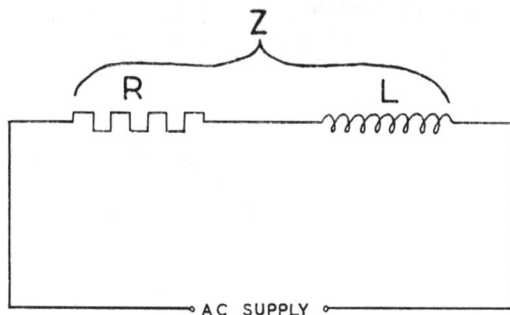

FIGURE 8.14   Example 8.3.

Solution

(a)
$$X_L = 2\pi f L$$
$$X_L = 2 \times \pi \times 50 \times 0{\cdot}05$$
$$X_L = 15{\cdot}71 \ \Omega$$
$$Z = \sqrt{R^2 + X_L{}^2}$$
$$Z = \sqrt{25^2 + 15{\cdot}71^2}$$
$$\underline{Z = 29{\cdot}5 \ \Omega}$$

(b)
$$I = \frac{E}{Z}$$
$$I = \frac{240}{29{\cdot}5}$$
$$\underline{I = 8{\cdot}13 \ A}$$

(c)   p.d. across $L = V_L = X_L I$
$$V_L = 15{\cdot}71 \times 8{\cdot}13$$
$$\underline{V_L = 128 \ V}$$
p.d. across $R = V_R = RI$
$$V_R = 25 \times 8{\cdot}13$$
$$\underline{V_R = 203 \ V}$$

(d)

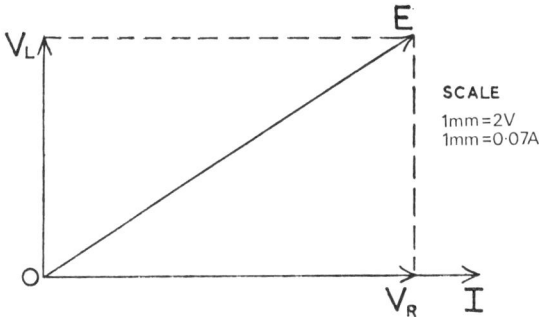

FIGURE 8.15   Phasor Diagram for Example 8.3.

8.84 The relation between R, X_L, and Z in a simple series circuit can be represented by drawing a right angled triangle with sides proportional to R, X_L and Z respectively as shown in Fig. 8.16. If this triangle is compared with the phasor diagram shown in Fig. 8.13 it can be seen that it is similar to the triangle *oab* and, in particular, that the angle ϕ is the same as the phase angle for the circuit. From the diagram it can be seen that:

$$R = Z \cos \phi$$
$$X_L = Z \sin \phi$$
$$\frac{X_L}{R} = \tan \phi$$

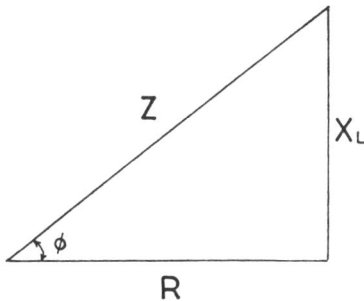

FIGURE 8.16   Impedance Triangle.

8.9. Power Used in a Single Phase Series Circuit

8.91   It was shown in Section 8.6 that the power consumed by a resistor can be found by using the formula:

$$P = RI^2$$

Also in Section 8.7 it was shown that a pure inductor consumes no power.

175

ELECTRICAL PRINCIPLES

Thus the total power consumed by the series circuit shown in Fig. 8.17 is:

$$P = RI^2 + 0$$
$$\therefore\quad P = RI \times I$$
$$\text{and since}\quad R = Z \cos \phi$$
$$P = ZI \cos \phi \times I$$
$$\text{but}\quad E = ZI \quad \text{(See 8.83)}.$$
$$\therefore\quad P = E \cos \phi \times I$$

Rewriting the formula

$$P = EI \cos \phi \text{ watts.}$$

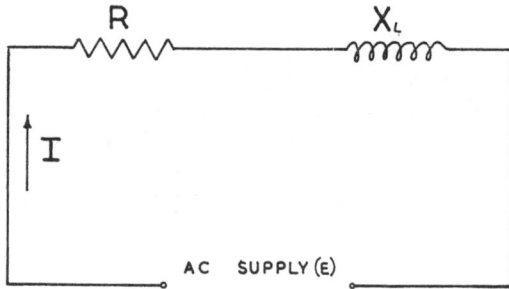

FIGURE 8.17   Series Circuit.

8.92   The above expression shows that in an a.c. circuit the power is not simply the product of electric pressure and current. The product $E \times I$ can be called the *apparent power* in the circuit: it is measured in *voltamperes* (abbreviation V A). The true power is measured in *watts* and is often less than the apparent power, although it can never be more.

The ratio

$$\frac{\text{watts}}{\text{volt amperes}}$$

is called the *power factor* (abbreviation p.f.) of the circuit.

$$\text{p.f.} = \frac{\text{watts}}{\text{volt amperes}} = \frac{\text{true power}}{\text{apparent power}}$$

For a single phase circuit:

$$\text{p.f.} = \frac{EI \cos \phi}{EI}$$

$$\therefore\quad \text{p.f.} = \cos \phi$$

8.93   If the power factor of a circuit is low the amount of current required to supply a given load can become excessive. For example, consider a load which consumes a power of 1 kW from a 100 V single phase a.c. supply:

The load current is given by:

$$I = \frac{P}{E \cos \phi}$$

At unity power factor: $I = \dfrac{1000}{100 \times 1} = 10 \text{ A}$

At 0·8 p.f.: $I = \dfrac{1000}{100 \times 0·8} = 12·5 \text{ A}$

At 0·4 p.f.: $I = \dfrac{1000}{100 \times 0·4} = 25 \text{ A}$

i.e. at 0·4 p.f. it would require a current of 25 A to supply the *same power* which could be supplied by only 10 A at unity p.f. It can be seen that the *lower* the power factor the *higher* will be the current required to supply a given load power. It follows that, if the power factor is low, it may be necessary to install larger capacity cables, transformers, switchgear, etc., than would be required for a unity p.f. load. There is also the possibility of higher voltage drops owing to the increased value of current in the supply cables.

8.94 The size and cost of alternating current plant is determined, to a large extent, by the electrical pressure and current which it has to supply. This is because the magnitude of the electric pressure determines the amount of insulation required and the magnitude of the current determines the cross-sectional area of the conductors. It is therefore usual to give the rating of a.c. plant in kV A rather than in kW. For example, consider a transformer which is designed to supply a pressure of 250 V and a maximum current of 40 A. The rating of this transformer is

$$\frac{250 \times 40}{1000} = 10 \text{ kVA}$$

If this transformer were used to supply a heating load at unity p.f., it could supply a maximum of $10 \times 1 = 10$ kW of load power. But if the same transformer were used to supply a motor load operating at 0·8 p.f. the maximum it could supply would be $10 \times 0·8 = 8$ kW of load power. It can be seen that by quoting the rating in kV A a figure is obtained which is independent of the load power factor for a given size transformer.

8.10   The Practical Inductor

8.101   A practical inductor, often called a choke or choking coil, consists essentially of a coil of insulated wire. The coil is often mounted on a laminated iron core to increase its inductance. Such coils can be used

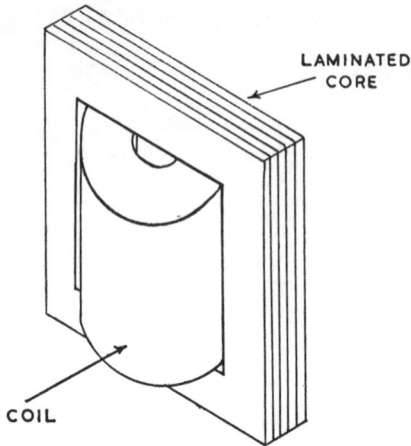

FIGURE 8.18 Practical Inductor.

to limit the current flowing in an a.c. circuit. They are more efficient for this purpose than resistors as they consume very little power; indeed a pure inductor would consume no power. An inductor can also be used to provide a voltage surge by suddenly interrupting the flow of current through it (see Section 7.6).

8.102 A good example of the use of an inductor is given by the simple fluorescent lamp circuit shown in Fig. 8.19.* When the switch S is closed current flows through the inductor L and the lamp electrodes so preheating the electrodes. When S is opened the current is suddenly interrupted and so the inductor produces a voltage surge which strikes the lamp. Current now flows through the mercury vapour gas in the lamp tube, the value of this current being determined by the choking effect of the inductor.

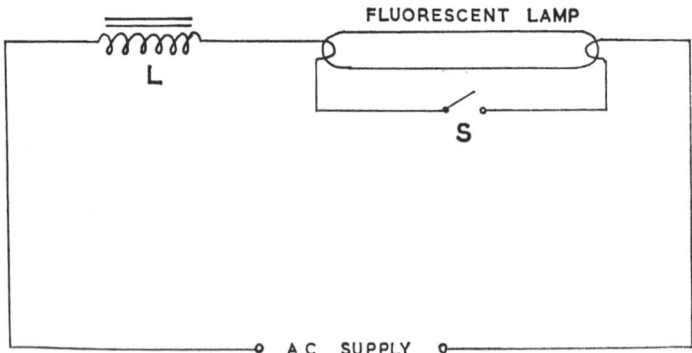

FIGURE 8.19 Fluorescent Lamp Circuit.

* A fuller account of the operation of fluorescent lamp circuits is given in the authors' "Electrical Installation Technology & Practice", E.U.P.

8.103 Although the main electrical property of a practical inductor is that of inductance, it must also inevitably possess resistance as the wire used has a resistance given by $R = \rho l/A$. In fact when iron cores are used the effective resistance is even higher than this value. It is convenient, for calculation purposes, to regard the practical inductor as consisting of a pure resistor and a pure inductor connected in series as shown in Fig. 8.20.

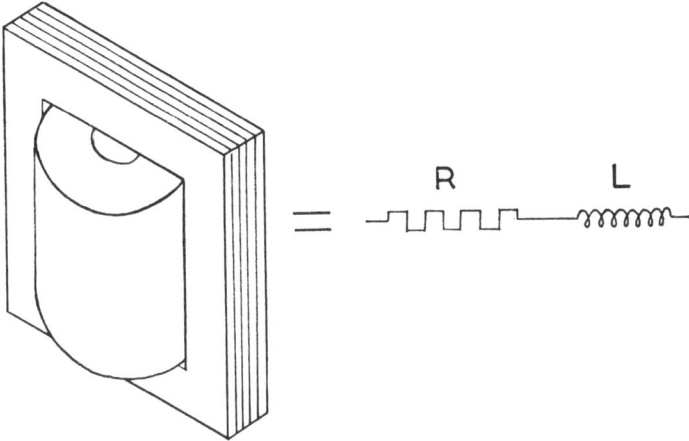

FIGURE 8.20 Equivalent Circuit of Practical Inductor.

Example 8.4

What is the function of a choking coil in an alternating current circuit?
When a d.c. supply at 240 V is applied to the ends of a certain choking coil, the current in the coil is 20 A. If an a.c. supply at 240 V, 50 Hz is applied to the coil, the current in the coil is 12·15 A.
Calculate the impedance, reactance, inductance and resistance of the coil.
What would be the general effect on the current if the frequency of the a.c. supply were increased?   C.G.L.I.

Solution

A choking coil can be used to limit the current in an a.c. circuit with a minimum loss of power. The choking coil is equivalent to a pure resistance in series with a pure inductance as shown in Fig. 8.21.
When d.c. is applied the current is limited only by the resistance of the coil:

$$\therefore \quad R = \frac{E}{I} \quad \text{(with d.c. supply)}$$

$$R = \frac{240}{20}$$

$$\underline{R = 12\,\Omega}$$

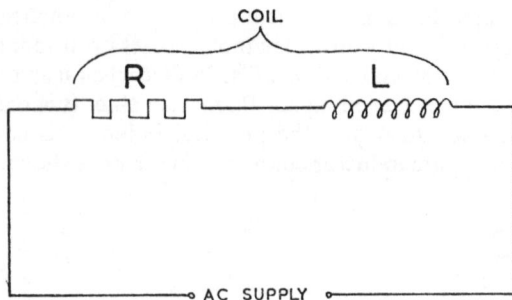

FIGURE 8.21   Example 8.4.

When a.c. is applied the current is limited by the impedance of the coil

$$\therefore \quad Z = \frac{E}{I} \quad \text{(with a.c. supply)}$$

$$Z = \frac{240}{12\cdot15}$$

$$\underline{Z = 19\cdot75\ \Omega} \qquad \text{say } 19\cdot8\ \Omega$$

$$X_L = \sqrt{Z^2 - R^2}$$

$$= \sqrt{19\cdot75^2 - 12^2}$$

$$\underline{X_L = 15\cdot7\ \Omega}$$

$$L = \frac{X_L}{2\pi f}$$

$$L = \frac{15\cdot7}{2 \times \pi \times 50}$$

$$\underline{L = 0\cdot05\ \text{H}}$$

If the frequency of the a.c. supply were increased the inductive reactance of the coil would increase. This is turn would increase the overall impedance of the circuit so causing a reduction in current.

8.11   Capacitance in A.C. Circuits

8.111   As was explained in paragraph 6.82 a capacitor is a device for storing an electric charge, the amount of charge stored for a given p.d. between the plates being given by:

$$Q = VC$$

Figure 8.22(a) shows a capacitor which has an alternating electric pressure applied to it. The charge on the plates is always proportional to the p.d. between the plates. Thus as this p.d. varies an electric current must flow either into or out of the capacitor in order to maintain the charge at its correct value. The greater the rate of change of the p.d., the greater will be the rate of flow of current. Figure 8.22(b) shows the waveforms of the p.d. charge and current. When an alternating electric pressure is applied to a capacitor, it can be seen that the charge is increasing most rapidly as the p.d. rises through zero, and that when the p.d. and charge are at a maximum their rate of change is zero. Hence the current wave leads the p.d. wave by 90°. If the frequency of the a.c. supply is raised the charge will have less time in which to flow into the capacitor, and therefore the rate of flow, or current, must increase. It follows that the current flowing through a capacitor is proportional to:

(a) the size of the capacitor
(b) the p.d. applied to the capacitor
(c) the frequency.

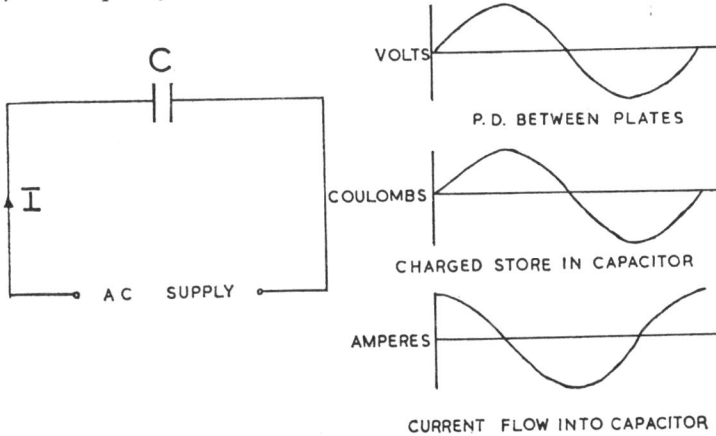

(a)   (b)
FIGURE 8.22   Capacitor in a.c. Circuit.
(a) Circuit Diagram   (b) Waveform.

8.112 The current passed by a pure capacitor when used in an a.c. circuit is given by:

$$I = E \times 2\pi fC$$

It follows that the reactance of the capacitor in ohms is given by:

$$X_C = \frac{E}{I} = \frac{1}{2\pi fC}$$

where: X_C = reactance of the capacitor in ohms
              f = frequency in Hz
              C = capacitance in farads.

In practice, capacitance values are often quoted in microfarads (abbreviation μF) and in this case it is often more convenient to write:

$$X_C = \frac{10^6}{2\pi fC}$$

C being measured in μF.

8.113. As explained in Section 8.111 above, the current in a pure capacitor leads the applied pressure by 90°. It follows that, as the phase angle is 90°, the power factor of a pure capacitor is cos 90° = 0; therefore a pure capacitor consumes no power. The table below compares the properties of inductance and capacitance in an a.c. circuit.

<table>
<tr><th colspan="3">Comparison of Inductance and Capacitance</th></tr>
<tr><th>Item</th><th>Inductance</th><th>Capacitance</th></tr>
<tr><td>Symbol</td><td>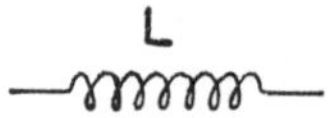</td><td>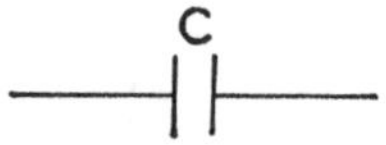</td></tr>
<tr><td>Reactance</td><td>$X_L = 2\pi fL$</td><td>$X_C = \frac{1}{2\pi fC}$</td></tr>
<tr><td>Phase relationship</td><td>Current lags by 90°</td><td>Current leads by 90°</td></tr>
<tr><td>Wave diagram</td><td>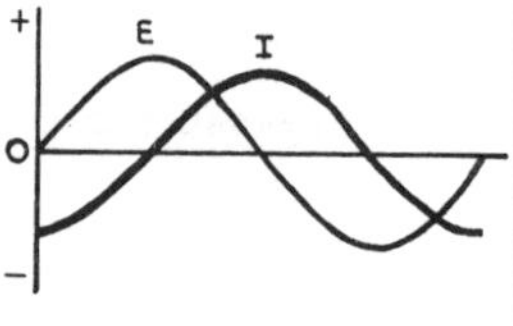</td><td>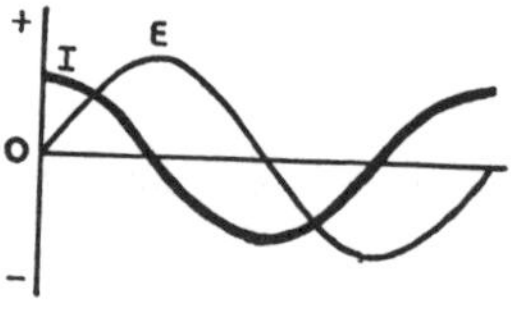</td></tr>
<tr><td>Phasor diagram</td><td>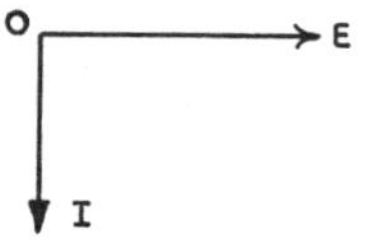</td><td>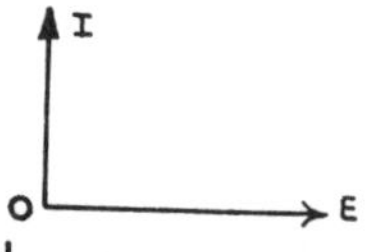</td></tr>
<tr><td>Variation of current with frequency</td><td></td><td>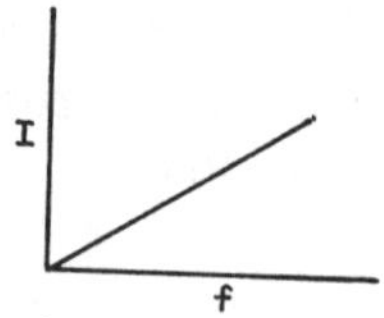</td></tr>
<tr><td>Variation of reactance with frequency</td><td>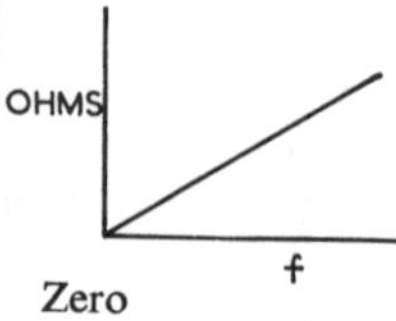</td><td>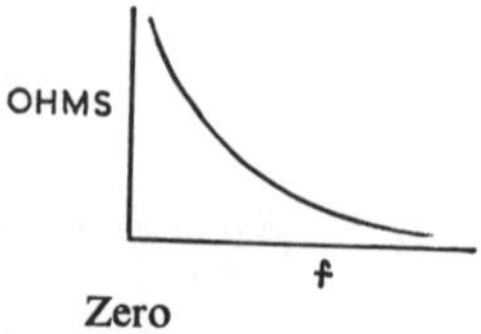</td></tr>
<tr><td>Power consumed</td><td>Zero</td><td>Zero</td></tr>
</table>

8.114 When a capacitor is connected in series with a resistor as shown in Fig. 8.23(a); the phasor diagram of the electric pressures is as shown in Fig. 8.23(b). It can be seen that:

$$E^2 = V_R{}^2 + V_C{}^2$$

As $\quad V_R = RI$

and $\quad V_C = X_C I$

where $X_C = 1/2\pi f C$.

It follows that:

$$\frac{E^2}{I^2} = R^2 + X_C{}^2$$

Hence

$$Z^2 = R^2 + X_C{}^2$$
$$Z = \sqrt{R^2 + X_C{}^2}$$

Also the power factor is:

$$\cos\phi = \frac{R}{Z} \quad \text{and is leading.}$$

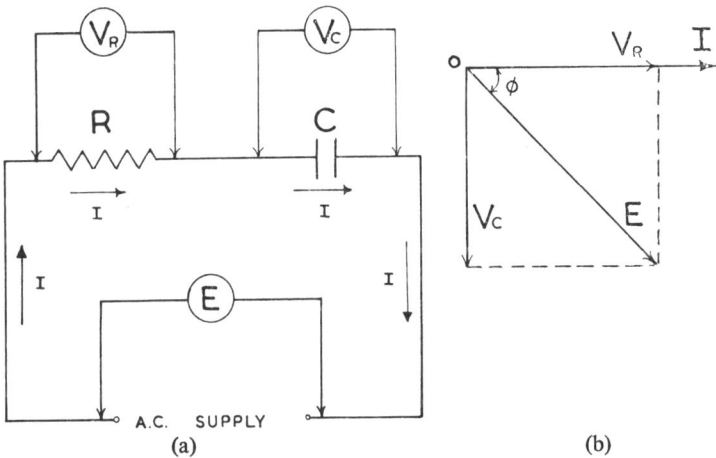

FIGURE 8.23   Resistance and Capacitance in Series.
(a) Circuit diagram.
(b) Phasor diagram.

Example 8.5

A 31·9 μF capacitor is connected in series with a 60 Ω non-inductive resistor and the combination connected to a 240 V, 50 Hz supply. Calculate

(a) the reactance of the capacitor
(b) the impedance of the circuit
(c) the current consumed by the circuit
(d) the power consumed by the circuit and the power factor.

183

Solution

(a)
$$X_C = \frac{1}{2\pi f C}$$

$$X_C = \frac{10^6}{2 \times \pi \times 50 \times 32}$$

$$\underline{X_C = 100 \; \Omega}$$

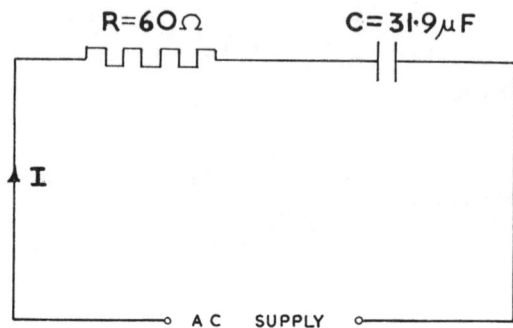

FIGURE 8.24   Example 8.5.

(b)
$$Z = \sqrt{R^2 + X_C^2}$$

$$Z = \sqrt{60^2 + 100^2}$$

$$\underline{Z = 117 \; \Omega}$$

(c)
$$I = \frac{E}{Z}$$

$$I = \frac{240}{117}$$

$$\underline{I = 2 \cdot 06 \; A}$$

(d)
$$\text{p.f.} = \cos \phi = \frac{R}{Z}$$

$$\text{p.f.} = \frac{60}{117}$$

$$\underline{\text{p.f.} = 0 \cdot 515 \qquad \text{leading}}$$

$$P = EI \cos \phi$$

$$P = 240 \times 2 \cdot 06 \times 0 \cdot 515$$

$$\underline{P = 255 \; W}$$

(A good alternative method is to use $P = RI^2$ followed by p.f. $= P/EI$.)

8.12 General Series Circuit

8.121 Figure 8.25 shows a series circuit which contains resistance, inductance and capacitance. Such a circuit is often called a general series circuit. As the components of the circuit are connected in series they will

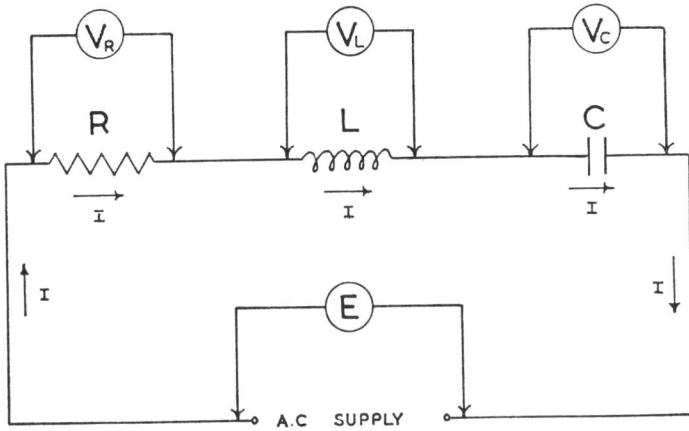

FIGURE 8.25   General Series Circuit.

all carry the same current, (I) and the supply e.m.f., (E) will be the phasor sum of the p.ds., V_R, V_L and V_C. The phasor diagram for the circuit is shown in Fig. 8.26 and it should be noted that:

(a) As the current is the same throughout a series circuit, I is chosen as the reference phasor.
(b) V_R is in phase with I.
(c) V_L leads I by 90°.
(d) V_C lags I by 90°.
(e) As V_L and V_C differ in phase by 180° their resultant is $V_L - V_C$.
(f) E is found by completing the parallelogram with sides given by this resultant $(V_L - V_C)$ and V_R.

8.122   From the phasor diagram, Fig. 8.26, it can be seen that:

$$E^2 = V_R{}^2 + (V_L - V_C)^2$$

and so by dividing throughout by I^2:

$$\left(\frac{E}{I}\right)^2 = \left(\frac{V_R}{I}\right)^2 + \left(\frac{V_L}{I} - \frac{V_C}{I}\right)^2$$

$$Z^2 = R^2 + (X_L - X_C)^2$$

or

$$Z = \sqrt{R^2 + (X_L - X_C)^2}$$

185

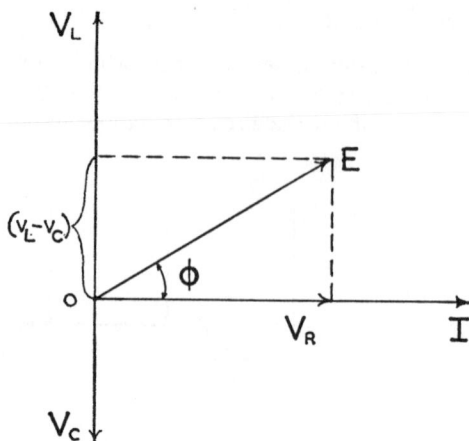

FIGURE 8.26   Phasor Diagram for General Series Circuit.

It is convenient for calculation purposes to let

$$X = X_L - X_C$$

so that $Z = \sqrt{R^2 + X^2}$

It should be noted that the sign of X can be positive or negative depending on which of the two reactances, X_L or X_C, is the greater. However X^2 is always positive. If X is positive this denotes a predominately inductive circuit and a lagging power factor, while if X is negative this denotes a predominately capacitive circuit and a leading power factor.

Example 8.6

What is meant by the following terms used in connection with alternating current; resistance, impedance and reactance?

A pressure of 240 V, at a frequency of 50 Hz is applied to the ends of a circuit containing a resistance of 5 ohms, an inductance of 0·02 H and a capacitance of 150 microfarads, all in series. Calculate the current in the circuit.

C.G.L.I.

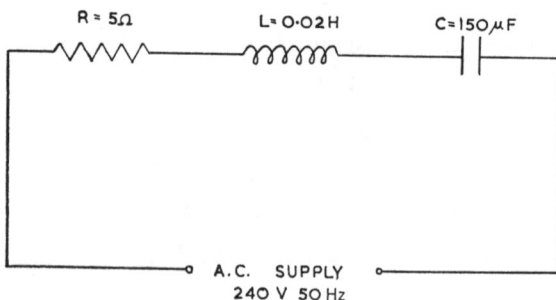

FIGURE 8.27   Example 8.6.

Solution

Resistance is the opposition afforded by a pure resistor to the passage of electric current.

Impedance is the total opposition afforded by a circuit to the passage of alternating electric current.

Reactance is the opposition afforded by a pure inductor or pure capacitor to the passage of alternating current.

Resistance, impedance and reactance are all measured in ohms.

$$X_L = 2\pi f L$$
$$X_L = 2 \times \pi \times 50 \times 0.02$$
$$X_L = 6.28 \ \Omega$$
$$X_C = \frac{1}{2\pi f C}$$
$$X_C = \frac{10^6}{2\pi \times 50 \times 150}$$
$$X_C = 21.21 \ \Omega$$
$$X = X_L - X_C$$
$$X = 6.28 - 21.21$$
$$X = -14.93 \ \Omega$$

(The minus sign indicates that the circuit is predominantly capacitive, and hence the overall p.f. is leading.)

$$Z = \sqrt{R^2 + X^2}$$
$$Z = \sqrt{5^2 + 14.93^2}$$
$$Z = 15.75 \ \Omega$$
$$I = \frac{E}{Z}$$
$$I = \frac{240}{15.75}$$
$$\underline{I = 15.2 \text{ A}}$$

8.123 If the frequency of the supply e.m.f. applied to the general series circuit of Fig. 8.25 is varied from a low to a high value, the inductive reactance X_L increases and the capacitive reactance X_C decreases. Figure 8.28 shows how X_L and X_C vary with frequency and it can be seen that at some frequency $X_L = X_C$. The frequency at which $X_L = X_C$ is called the *resonant frequency* (symbol f_r) of the circuit and when this frequency is applied to the circuit:

$$Z = \sqrt{R^2 + (X_L - X_C)^2}$$

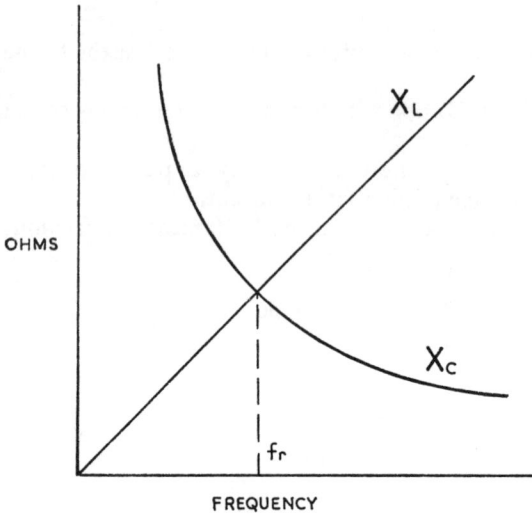
FIGURE 8.28   Variation of X_L and X_C with Frequency.

But as
$$X_L = X_C$$
$$X_L - X_C = 0$$
Therefore
$$Z = \sqrt{R^2 + 0^2}$$
i.e.
$$Z = R$$

The frequency at which resonance occurs can be found as follows:

As
$$X_L = X_C$$
it follows that:
$$2\pi f_r L = \frac{1}{2\pi f_r C}$$
$$(2\pi f_r)^2 = \frac{1}{LC}$$
$$2\pi f_r = \frac{1}{\sqrt{LC}}$$
$$f_r = \frac{1}{2\pi\sqrt{LC}}$$

At resonance the impedance of the circuit is simply equal to its resistance. At any other frequency the resultant value of X is not zero and so the impedance is higher than R. The power factor of any series circuit is given by

$$\cos \phi = \frac{R}{Z}$$

188

but as at resonance $Z = R$, it follows that:

$$\text{p.f.} = \cos \phi = \frac{R}{R} = 1$$

i.e. the power factor at resonance is unity.

The current flowing at resonance is given by $I = E/R$ and this value of current flows through both reactances X_L and X_C, giving rise to p.ds. of value $X_L I$ and $X_C I$. These p.ds. are equal to each other and can have a value very much higher than the supply voltage. Figure 8.29 shows how the voltages across the inductor and capacitor of a general series circuit

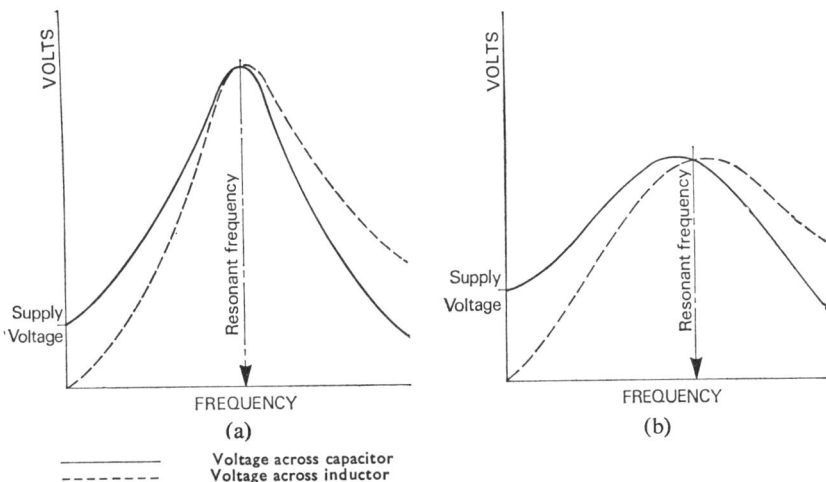

Voltage across capacitor
Voltage across inductor

FIGURE 8.29   Resonance Curves.
(a) Low resistance circuit.
(b) High resistance circuit.

vary as the frequency is varied through resonance. It can be seen that in a low resistance circuit the supply voltage is "magnified" by the resonance effect to a far greater extent than in a high resistance circuit. Also the curves for a low resistance circuit are sharper or more "selective" than they are for a high resistance circuit.

Example 8.7

A coil of resistance 20 Ω and inductance 0·08 H is connected to a supply at 240 V, 50 Hz. Calculate:

(a) the current in the circuit
(b) the value of a capacitance to be put in series with the coil so that the current shall be 12 A.   C.G.L.I.

189

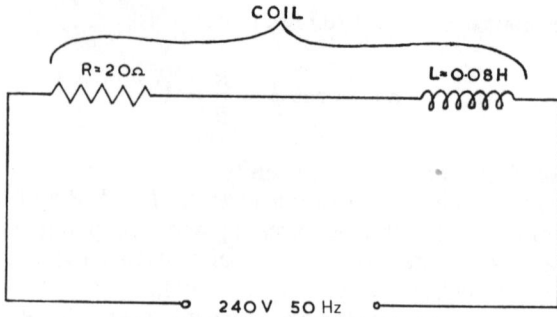

FIGURE 8.30   Example 8.7 part (a).

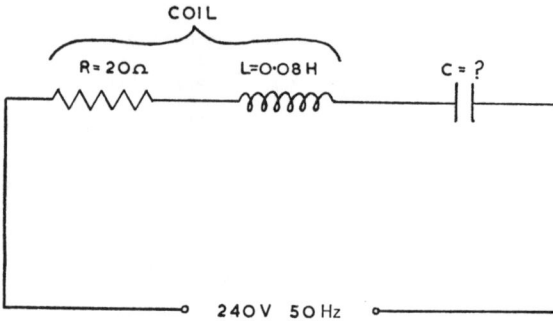

FIGURE 8.31   Example 8.7 part (b).

Solution

(a)

$$X_L = 2\pi f L$$
$$X_L = 2 \times \pi \times 50 \times 0{\cdot}08$$
$$X_L = 25{\cdot}14 \ \Omega$$
$$Z = \sqrt{R^2 + X_L{}^2}$$
$$Z = \sqrt{20^2 + 25{\cdot}14^2}$$
$$Z = 32{\cdot}1 \ \Omega$$
$$I = \frac{E}{Z}$$
$$I = \frac{240}{32{\cdot}1}$$
$$\underline{I = 7{\cdot}48 \ \text{A}}$$

(b)

$$Z = \frac{E}{I}$$
$$Z = \frac{240}{12}$$
$$Z = 20 \ \Omega$$

As in this case $Z = R$ it follows that the circuit is resonant and therefore $X_L = X_C$

$$X_C = X_L = 25\cdot14\ \Omega$$

$$\frac{1}{2\pi f C} = 25\cdot14$$

$$C = \frac{1}{2\pi \times 50 \times 25\cdot14}$$

$$C = 0\cdot000\ 127\ F \quad \text{or} \quad 127\ \mu F$$

Example 8.8

A coil of inductance 0·05 H is connected in series with a 10 μF capacitor. At what frequency does resonance occur?

Solution

$$f_r = \frac{1}{2\pi\sqrt{LC}}$$

$$f_r = \frac{1}{2\pi\sqrt{0\cdot05 \times 10 \times 10^{-6}}}$$

Note that the factor 10^{-6} is required to convert the value of capacitance quoted in μF into a value in Farads.

$$f_r = 225\ Hz$$

SUMMARY OF FORMULAE

Periodic time
$$T = \frac{1}{f}$$

Values of a sine wave:
$\begin{cases} \text{average value} = 0\cdot636 \times \text{maximum value} \\ \text{r.m.s. value} = 0\cdot707 \times \text{maximum value} \\ \text{form factor} = \dfrac{\text{r.m.s.}}{\text{average}} = 1\cdot11 \end{cases}$

Average power dissipated:
$\begin{cases} \text{in a pure resistor} \quad P = RI^2 \\ \text{in a pure inductor} \quad P = 0 \\ \text{in a pure capacitor} \quad P = 0 \end{cases}$

Reactance of an inductor:
$$X_L = 2\pi f L$$

Reactance of a capacitor: $\quad X_C = \dfrac{1}{2\pi f C}$

Reactance of a series circuit: $\quad X = X_L - X_C$

Impedance of a series circuit: $\quad \begin{cases} Z = \dfrac{E}{I} \\[2mm] Z = \sqrt{R^2 + X^2} \end{cases}$

Power factor: $\quad \begin{cases} \text{p.f.} = \cos\phi \\[2mm] \cos\phi = \dfrac{R}{Z} \\[2mm] \cos\phi = \dfrac{P}{EI} \end{cases}$

Power in a series circuit: $\quad \begin{cases} P = RI^2 \\[2mm] P = EI\cos\phi \end{cases}$

Resonant frequency: $\quad f_r = \dfrac{1}{2\pi\sqrt{LC}}$

EXERCISE 8

1. (a) Explain the following terms applied to an alternating current wave:
    (i) maximum value,
    (ii) average value,
    (iii) r.m.s. value.
    If the maximum value of a sine wave is 300 A, give the average and r.m.s. values.
   (b) A single-phase motor takes 16 amperes r.m.s. at 240 volts r.m.s. Assuming that the current lags by 30° (electrical), calculate the power input to the motor.
   C.G.L.I.

2. Explain the terms r.m.s., average and maximum values of an alternating current of sine waveform and state the one normally used.
    Sketch the waveforms of voltage and current of an a.c. circuit having resistance and inductive reactance of equal values connected to a voltage of sine waveform.
   E.M.E.U.

3. Explain the term, "impedance of a circuit".
    A coil of inductance of 0·2 H and resistance 50 Ω is connected to a 200 V, 50 Hz, a.c. supply. Calculate:
   (a) the reactance of the coil,
   (b) the impedance of the coil,
   (c) the supply current, and
   (d) the phase angle between the voltage and current.
   E.M.E.U.

4. Explain the property of inductance of a coil and define the unit of inductance.
    A coil having resistance 50 Ω and inductance 0·2 H is connected to a 250 V, 50 Hz, a.c. supply. Determine the coil reactance, impedance and current.
   E.M.E.U.

5. A coil of resistance 20 Ω takes 4 A from a 240 V, 50 Hz, a.c. supply. Find (a) the impedance, (b) the reactance, (c) the inductance, (d) the phase angle.   E.M.E.U.

6. A coil of resistance 8 ohm and inductance 0·04 H is connected to a 240 V, 50 Hz. a.c. supply.
   Determine the coil reactance, impedance, current and the phase angle between the voltage and current. Draw the voltage and current phasors with the correct phase angle between them.   E.M.E.U.

7. (a) What do you understand by "choking coil"? Explain its operation and use.
   (b) The impedance at 50 Hz of a certain choking coil is 12·36 ohms, and the inductance is 0·03 henry. Calculate:
      (i) the resistance of the coil,
      (ii) the current when a d.c. supply at 240 V is applied to the ends of the coil.
      (iii) the current when an a.c. supply at 240 V, 50 Hz is applied to the ends of the coil.   C.G.L.I.

8. Explain the term "choking coil". For what purpose is a choking coil used?
   A choking coil takes 8 amperes when connected to a d.c. supply at 240 volts. If the coil is connected to an a.c. supply at 240 volts, 50 hertz, the current is 4·8 amperes, calculate: (a) the resistance, (b) the inductance, and (c) the impedance, of the coil.   C.G.L.I.

9. A single-phase alternating current supply at 240 V, 50 Hz is applied to a series circuit consisting of an inductive coil of negligible resistance, and a non-inductive resistance coil of 15 ohms. When a voltmeter is applied to the ends of each coil in turn, the potential differences are found to be: 127·5 V across inductive coil, 203 V across resistance. Calculate:
   (a) the impedance of the circuit,
   (b) the inductance of the coil,
   (c) the current in the circuit and
   (d) the power factor.   C.G.L.I.

10. Define the term "power factor", and state how it affects cable size.
   A choking coil of resistance 8 ohms and of inductance 0·015 henry, is connected to an alternating current supply at 240 volts, single-phase, 50 hertz. Calculate:
   (a) the current from the supply,
   (b) the power in the circuit,
   (c) the power factor.   C.G.L.I.

11. Explain the meaning of the following terms used in connection with alternating current: *inductance, capacitance, reactance* and *impedance*.
   A circuit comprises a resistance of 6 ohms, an inductance of 0·05 henry and a capacitance of 320 microfarads, all connected in series. An alternating supply at 240 V 50 Hz is applied to the ends of the circuit. Calculate the current taken.   C.G.L.I.

12. (a) A pressure of E volts at a frequency of f hertz is applied to the ends of a series circuit containing resistance R ohms, inductance L henry, and capacitance C farads. Write down a formula which would give the current I amperes in terms of the above quantities.
   (b) An a.c. circuit consisting of a resistance 5 ohms in series with an inductance is supplied with pressure at 240 V, 50 hertz. If the current in the circuit is 25 amperes, calculate the value of the inductance in henrys.   C.G.L.I.

9 Alternating Currents II

9.1 Components of Phasors

9.11 Any phasor quantity may be considered as being the phasor sum of two components. In dealing with a current flowing in an electric circuit the components which are of particular importance are:

(a) the in-phase, or active component,
(b) the quadrature, or re-active component.

Consider a current flowing in an electric circuit which lags the supply e.m.f. by a phase angle ϕ, as shown in Fig. 9.1. The current may be considered as being composed of an active component, I active, which is in phase with E, and a reactive component, I reactive, which lags E by 90°. It can be seen from the phasor diagram, Fig. 9.1 that:

$$I \text{ active} = I \cos \phi$$
$$I \text{ reactive} = I \sin \phi$$
$$I = \sqrt{I^2 \text{ active} + I^2 \text{ reactive}}$$

A component of a phasor can take either of two directions, for example in Fig. 9.1(a) the reactive component is lagging the reference phasor E and

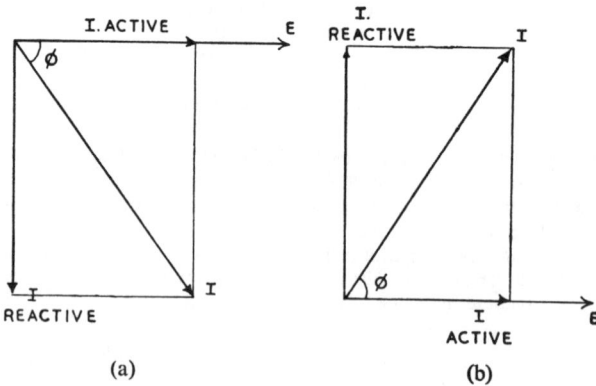

FIGURE 9.1 Components of a Current.
(a) Lagging Current.
(b) Leading Current.

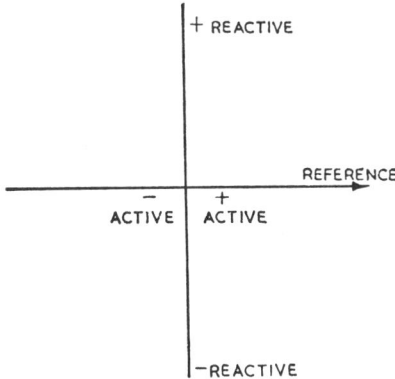

FIGURE 9.2 Sign Convention for Components.

points downwards, while in 9.1(b) the reactive component is leading the reference phasor E and points upwards. This means that a sign convention is necessary to distinguish between the two possibilities. The convention normally used is illustrated in Fig. 9.2. It should be noted that, in practice, the active component is almost always positive, and the reactive component is *positive* for a *leading* current and *negative* for a *lagging* current.

9.12 The use of components simplifies the process of phasor addition. Two phasors cannot be added arithmetically unless they happen to be in phase with each other, but as the *active* components of the two phasors *are in phase* with each other they can be added numerically to obtain the total active component. Similarly the *reactive* components are either *in phase* or *180° out of phase* with each other and so the resultant reactive component is the *algebraic* sum of the individual reactive components, i.e. due account must be taken of the signs of the components when adding.

Example 9.1

A load which consumes a current of 10 A at 0·8 p.f. leading is connected in parallel with another load which consumes 12 A at 0·5 p.f. lagging.

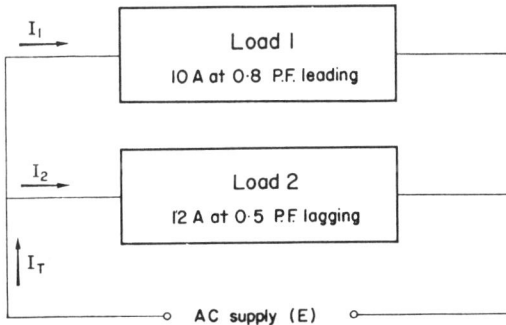

FIGURE 9.3 Circuit Diagram for Example 9.1.

195

Calculate the overall current taken from the supply and the overall power factor. Check your answer by means of a phasor diagram drawn to scale.

Solution

As the two loads are connected in parallel, I is the phasor sum of I_1 and I_2. This phasor sum can be calculated by using components. It is convenient to set out the calculations in tabular form as below:

Current	Active Components	Reactive Components
I_1	$I_1 \cos \phi_1 = 10 \times 0.8 = +8$ A	$I_1 \sin \phi_1$ (leading) $= 10 \times 0.6 = +6$ A
I_2	$I_2 \cos \phi_2 = 12 \times 0.5 = +6$ A	$I_2 \sin \phi_2$ (lagging) $= 12 \times 0.866 = -10.39$ A
I_T	$+14$ A	-4.39 A

It should be noted that the overall current I_T has a lagging power factor as its reactive component is negative

$$I_T = \sqrt{I^2 \text{ active} + I^2 \text{ reactive}}$$

$$I_T = \sqrt{14^2 + 4.39^2}$$

$$\underline{I_T = 14.7 \text{ A}}$$

As I active $= I_T \cos \phi_T$

$$\text{Overall p.f.} = \cos \phi_T = \frac{I \text{ active}}{I_T}$$

$$\cos \phi_T = \frac{14}{14.7}$$

$$\underline{\cos \phi_T = 0.95 \text{ lagging}}$$

9.13　The use of components is not confined to the addition of currents as above. For example the p.ds. across two loads connected in series may be resolved with respect to their common current phasor, in order to calculate the resultant total electric pressure. In this case

$$V \text{ active} = V \cos \phi$$

$$V \text{ reactive} = V \sin \phi$$

Thus　　$$V = \sqrt{V^2 \text{ active} + V^2 \text{ reactive}}$$

The components of an impedance, Z, can be regarded as a resistance and a reactance connected in series. In this case R is the active, or power

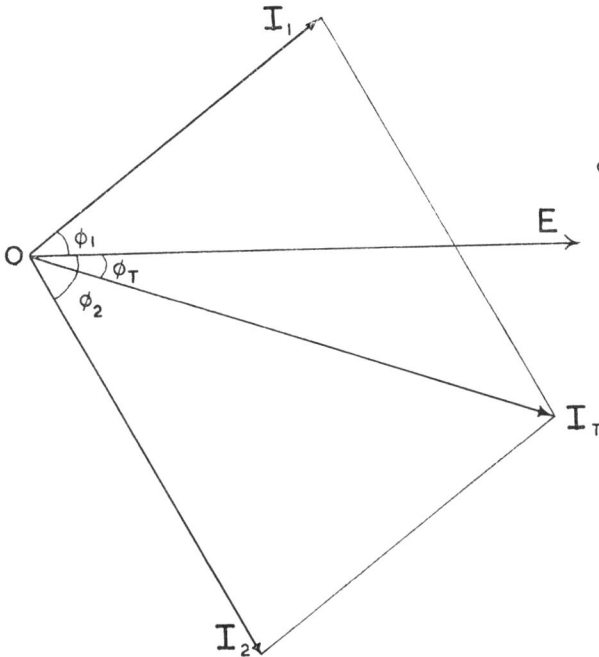

FROM THE
DIAGRAM

$I_T = 14 \cdot 7$ A.

$\phi_T = 18 \cdot 2^{\circ}$ OR $18^{\circ} 12'$

COS. $\phi_T = 0 \cdot 95$
LAGGING

FIGURE 9.4   Phasor Diagram for Example 9.1.

consuming component and X is the reactive, or non-power consuming component.

$$\text{Active component} = R = Z \cos \phi$$

$$\text{Reactive component} = X = Z \sin \phi$$

$$Z = \sqrt{R^2 + X^2}$$

9.14 The apparent power in an a.c. circuit, which is measured in volt amperes, can also be resolved into two components at right angles to each other. Fig. 9.5(a) shows the phasor diagram for a circuit carrying a lagging current. The components of this current are:

$$I \text{ active} = I \cos \phi$$

$$I \text{ reactive} = I \sin \phi$$

Each current shown in the phasor diagram can be multiplied by the supply e.m.f. E so obtaining a volt ampere diagram Fig. 9.5(b). From the diagrams it can be seen that:

$$P = E \times I \text{ active} = EI \cos \phi$$

197

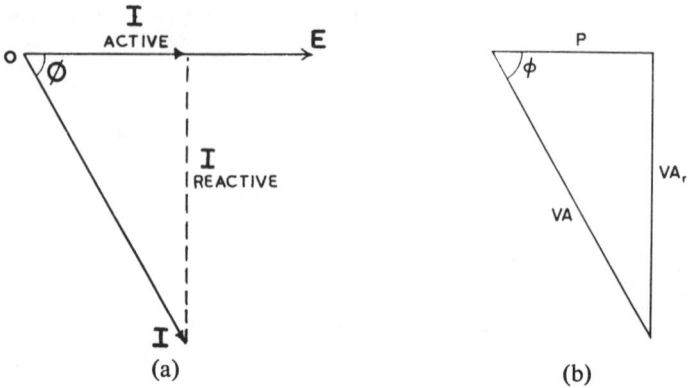

FIGURE 9.5   Phasor Diagram for Circuit with a Lagging Current.
(a) Components of current.
(b) Volt ampere diagram.

Thus the side P of the diagram represents the true power, P, consumed by the circuit, measured in watts.

Also $V Ar = E \times I \text{ reactive} = EI \sin \phi$

This component is called the volt amperes reactive (abbreviation V Ar) and is sometimes referred to as the "idle" or "wattless" component of the apparent power.

Example 9.2

 (a) What is meant by power factor?
 (b) The installation in a factory carries the following loads: Lighting 50 kW, heating 30 kW and power 51·45 kW. Assuming that the lighting and heating loads are non-inductive, and the power has an overall power factor of 0·7 lagging, calculate:
   (i) the total loading in kW;
   (ii) the kVA demand at full load.   C.G.L.I.

Solution

 (a) The power consumed by an a.c. circuit may be less than the product of the electric current and pressure because these two quantities may not be in phase with each other. The ratio of the true power, in watts, to the apparent power, in volt amperes, is called the power factor.

$$\text{Power factor} = \frac{P}{EI}$$

It can be shown that in a single phase a.c. circuit the power factor is equal to the cosine of the phase angle.

 (b) As the lighting and heating loads are non-inductive they have no reactive component.

The power load consumes:

$$P = 51 \cdot 45 \text{ kW}$$

$$\text{kVA} = \frac{\text{kW}}{\cos \phi}$$

$$= \frac{51 \cdot 45}{0 \cdot 7}$$

$$= 73 \cdot 5 \text{ kVA}$$

$$\text{kVAr} = \text{kV A} \times \sin \phi$$

$$= 73 \cdot 5 \times 0 \cdot 714 \, 2$$

$$= 52 \cdot 49 \text{ kVAr}$$

Load	Active Component (kW)	Reactive Component (kV Ar)
Lighting	50 kW	0 kVAr
Heating	30 kW	0 kVAr
Power	51·45 kW	−52·49 kVAr
Totals	131·45 kW	−52·49 kVAr

Note that a minus sign is assigned to the reactive component of power load to denote a lagging power factor.

(i) from the above table the total loading $= \underline{131 \text{ kW}}$

(ii) $\text{kV A} = \sqrt{\text{kW}^2 + \text{kV Ar}^2}$

$\quad \text{kV A} = \sqrt{131 \cdot 5^2 + 52 \cdot 5^2}$

$\quad \underline{\text{kV A} = 142}$

Figure 9.6 shows the complete volt ampere diagram for this example.

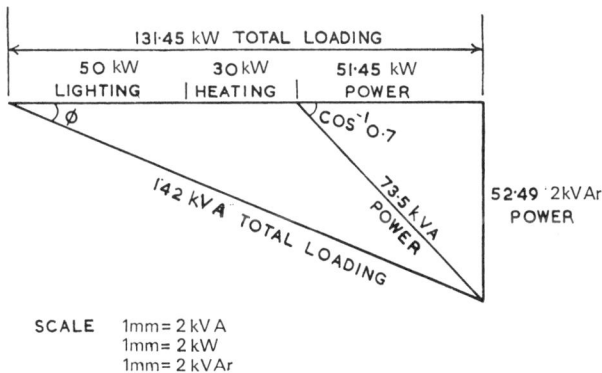

FIGURE 9.6   Example 9.2.

9.2 Power Factor Correction

9.21 As has been previously explained in paragraph 8.93 a low power factor is undesirable. In practice low power factors are often caused by inductive loads. These loads, which include equipment such as induction motors, inductor operated discharge lamps etc., operate at a low lagging power factor because of their inherent electrical characteristics. One method of improving the overall power factor of an installation is to connect a load having a leading power factor in parallel with the loads having lagging power factors. In this way although each load operates at its own power factor, the overall power factor of the combined load is improved. In large installations it may be possible to provide the leading load by installing a type of motor which can operate at a leading power factor, even though motors of this type are usually more expensive and require more elaborate control equipment than the simple induction motor. For example, in many collieries large synchronous motors, which can be operated at a leading power factor, are used to drive ventilating fans and at the same time improve the overall power factor of the complete installation. A pure capacitor provides a load which operates at zero power factor leading; thus, by connecting a capacitor in parallel with an inductive load, the overall power factor can be raised without any increase in the total true power consumed, and without any alteration in the current consumed by the load itself.

9.22 Consider an inductive load taking a current I_L lagging by a phase angle ϕ, from a supply of e.m.f. (E). To improve the overall power factor a capacitor C taking a current I_C can be connected in parallel with the load as shown in Fig. 9.7(a). The phasor sum of the capacitor current I_C and the

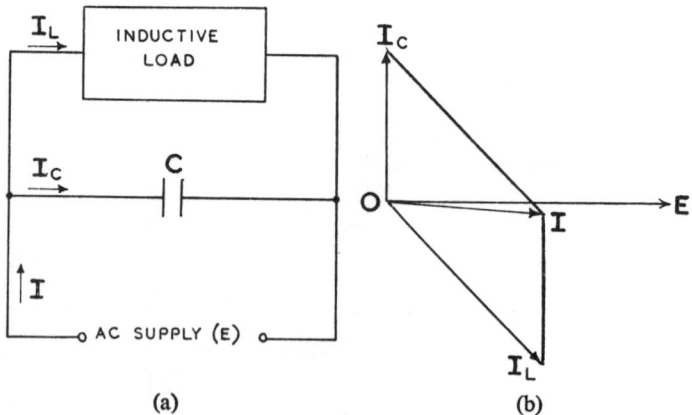

(a)   (b)

FIGURE 9.7 Power Factor Correction by Capacitor.
(a) Circuit diagram.
(b) Phasor diagram.

load current I_L produces a resultant current I which is at a smaller phase angle and hence a better power factor than the original load current I_L. This is illustrated by the phasor diagram shown in Fig. 9.7(b).

Example 9.3

An induction motor takes a current of 10 A at 0·7 p.f. lagging from a 240 V 50 Hz single phase a.c. supply. What value capacitor connected in parallel with the motor is required to improve the overall power factor to 0·95 p.f. lagging?

Solution

When dealing with problems of this type, the first step is to determine the required value of capacitor current. This can be easily determined

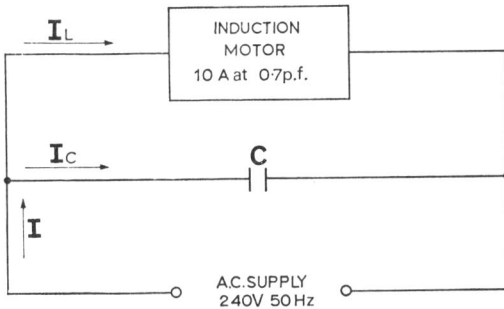

FIGURE 9.8   Example 9.3.

either by drawing a phasor diagram to scale or by using the component method of calculation.

(a) Phasor diagram method:

First determine the phase angles as follows:

Phase angle of motor current = $\cos^{-1} 0·7$ = 45° 35′ lagging
Phase angle of overall current = $\cos^{-1} 0·95$ = 18° 12′ lagging
Phase angle of capacitor current = 90° leading.

Step 1. Draw the supply e.m.f. as reference phasor and the motor current phasor at its correct phase angle as shown in Fig. 9.9.

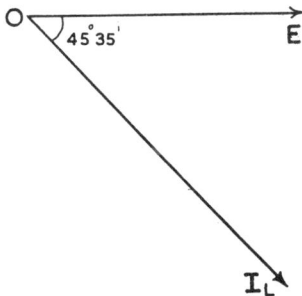

FIGURE 9.9   Step 1 in Drawing Phasor Diagram for Example 9.3.

201

Step 2. Draw a line *ab* from the end of the I_L phasor and at 90° to the e.m.f. phasor, and also a line *oc* at the required final phase angle. These lines intersect at point *d*, as shown in Fig. 9.10.

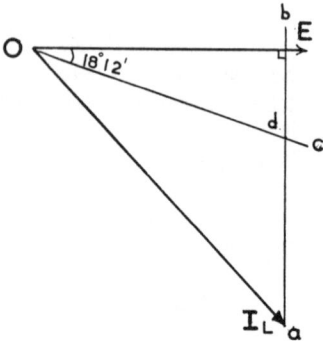

FIGURE 9.10 Step 2 in Drawing Phasor Diagram for Example 9.3.

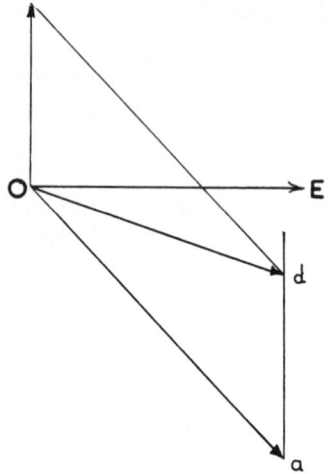

FIGURE 9.11 Step 3 in Drawing Phasor Diagram for Example 9.3.

Step 3. The phasor diagram can now be completed as *od* is the phasor representing the overall current, *I*, and *ad* represents the capacitor current, I_C. The phasor representing the capacitor current can be transferred to the origin by completing the parallelogram, as shown in Fig. 9.11. The value of I_C can now be measured from the phasor diagram and is found to be 4·85 A. Fig. 9.12 shows the complete phasor diagram drawn to scale.

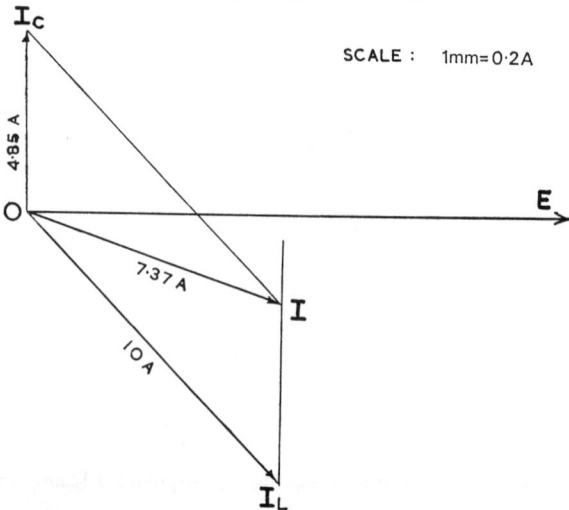

FIGURE 9.12 Completed Phasor Diagram for Example 9.3.

(b) Calculation by components:

It should be noted that:

  (i) The total current I is the phasor sum of I_L and I_C, thus I_C can be found by subtracting I_L from I using phasors.

  (ii) The active and reactive components of I_L can readily be found, using I_L active $= I_L \cos \phi_L$ and I_L reactive $= I_L \sin \phi_L$.

  (iii) The capacitor current has no active component and so the active component of I equals the active component of I_L. This enables I to be found,

$$I = \frac{I \text{ active}}{\cos \phi}$$

and from this I reactive $= I \sin \phi$ is determined.

Current	Active Component	Reactive Component
$I = \dfrac{7}{0.95} = 7.368$ A $I_L = 10$ A	7 A $10 \times 0.7 = 7$ A	$7.368 \times 0.3123 = -2.296$ A $10 \times 0.7143 = -7.143$ A
$I_C = I - I_L$ (phasor subtraction)	0 A	$+4.847$ A

(c) Now that I_C has been found, the reactance X_C can be evaluated and from this the value of capacitor C determined.

$$X_C = \frac{E}{I_C}$$

i.e.

$$X_C = \frac{240}{4.847}$$

Whence:

$$X_C = 49.52 \; \Omega$$

$$C = \frac{1}{2\pi f X_C}$$

i.e. substituting for X_C, we have:

$$C = \frac{1}{2\pi \times 50 \times 49.52}$$

i.e.

$$C = 64.3 \times 10^{-6} \text{ F}$$

Whence:

$$C = 64.3 \; \mu\text{F}$$

9.23 When the power factor of a number of induction motors connected to the same supply has to be improved by using capacitors, a decision has to be made concerning the location of the capacitors. It is possible to provide each motor with its own individual capacitor, alternatively one large capacitor bank can be used to provide overall power factor correction for the whole group of motors. When using individual capacitors the full load current in the leads to the motor is reduced to a minimum, and furthermore when the motor is switched off, the capacitor is also automatically switched off. If however the motor is running lightly loaded, the capacitor may well be too large and give an overall leading power factor. The cost of installing individual capacitors is often greater than that of installing a single large capacitor bank. When installing a single capacitor bank, this has to correct the combined power factor of all the motors in use at one time. This method has the advantage that it often requires less total capacitance than when using individual capacitors, as not all motors will be operating at full load at the same time and so some allowance may be made for diversity. The disadvantage is that unless special switching is used, the whole capacitor bank will be in circuit even when only a few motors are operating and this could result in a very low leading power factor.

Example 9.4

The total current taken by a number of identical single-phase motors is 200 A with a power factor of 0·5 when its capacitor is out of circuit. Calculate: (a) What the total current will be when the power factor is brought up to unity, by switching in the capacitor; and also, (b) What both the total current and the power factor will be when 50% of the motors are switched off with the capacitor still in circuit. Illustrate the above calculations by means of a phasor diagram for each of the three conditions concerned.   C.G.L.I.

Solution

The calculations are readily performed using components as below:

Current	Active Component	Reactive Component
I motor = 200 A	$I \cos \phi = 200 \times 0.5$ = 100 A	$I \sin \phi = 200 \times 0.866$ = −173·2 A

(a) As the p.f. is to be raised to unity, the capacitor current will exactly equal the reactive motor current,

i.e. $I_C = +173.2$ A reactive

and the resultant total current will simply be equal to the active component of the motor current,

i.e. $I_T = 100$ A

(b) The motor current is halved but the capacitor current is unchanged, therefore:

Current	Active Component	Reactive Component
I motor	50 A	−86·6 A
I_C	0 A	+173·2 A
I_T	50 A	+86·6 A

$$I_T = \sqrt{I^2 \text{ active} + I^2 \text{ reactive}}$$

$$\therefore \ I_T = \sqrt{50^2 + 86\cdot6^2}$$

$$\underline{I_T = 100 \text{ A}}$$

$$\text{Overall p.f.} = \cos \phi_2 = \frac{I \text{ active}}{I_r}$$

$$\cos \phi_T = \frac{50}{100}$$

$$\underline{\cos \phi_T = 0\cdot5 \text{ leading}}$$

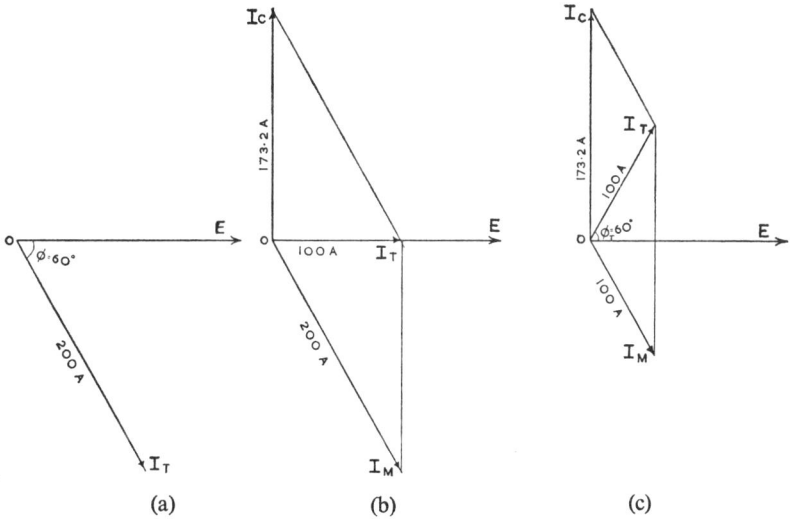

FIGURE 9.13 Phasor Diagrams for Example 9.4.
(a) Motor current with no capacitor in circuit.
(b) Conditions when p.f. is raised to unity.
(c) Condition with half motor current and full capacitor current.

9.24 If an inductor and capacitor are connected in parallel, the total current is minimum if the combined power factor is unity. The impedance of the combination is then a maximum and the circuit is said to exhibit parallel resonance. As in the case of series resonance, resonant conditions can be obtained by varying the capacitance the inductance or the frequency.

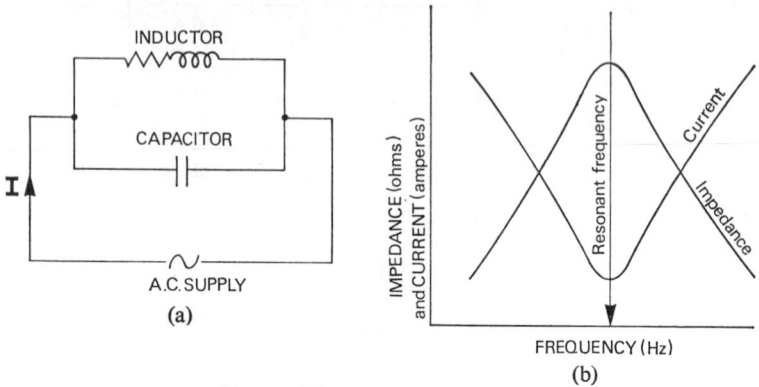

FIGURE 9.14   Parallel Resonance.
(a) Circuit.
(b) Variation of impedance and current with frequency.

Figure 9.14 shows how the impedance of a parallel resonant circuit varies with frequency.

9.3   Three Phase Supply Systems

9.31   An a.c. generator or alternator can be constructed by using a rotating magnetic field which induces e.m.fs. in three coils spaced at 120° intervals, as illustrated diagrammatically in Fig. 9.15. Each coil is called a *phase* and, as each phase coil has the same number of turns, each will have the same value of e.m.f. induced in it. However the e.m.fs. will have 120° phase displacement from each other as shown in Fig. 9.16. In order to distinguish between the phases they are called the *red*, *yellow* and *blue* phases respectively.

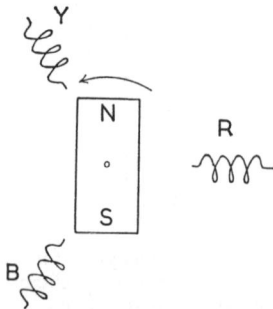

FIGURE 9.15   Principle of Three Phase Alternator.

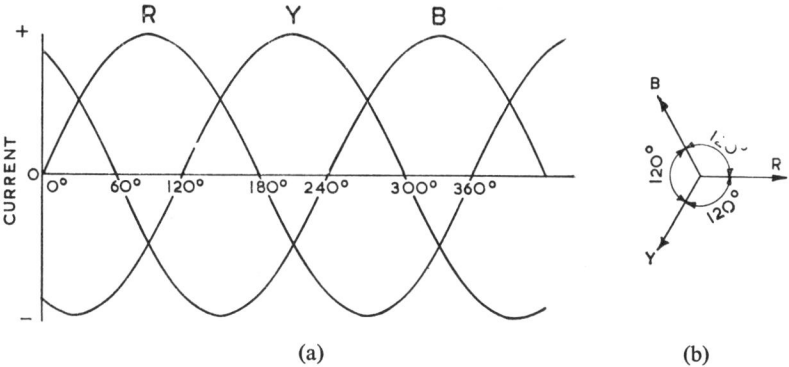

FIGURE 9.16   Three Phase e.m.f.s. (a) Wave diagram. (b) Phasor diagram.

9.32   There are two main methods of connecting the phases to produce a three-phase system. These are *star* connection and *delta* (or *mesh*) connection. The star connection is illustrated in Fig. 9.17 and it can be shown that when this method of connection is employed the electrical pressure between any two connecting lines (line voltage symbol, E_L) is $\sqrt{3}$ times the pressure developed in any one phase coil (phase voltage symbol E_P). The current in each phase coil (I_P) is the same as the current in the line connected to it (I_L)

$$E_L = \sqrt{3}\, E_P$$

$$I_L = I_P$$

The delta connection is illustrated in Fig. 9.18 and it can be shown that when this method of connection is employed the line voltage is equal to

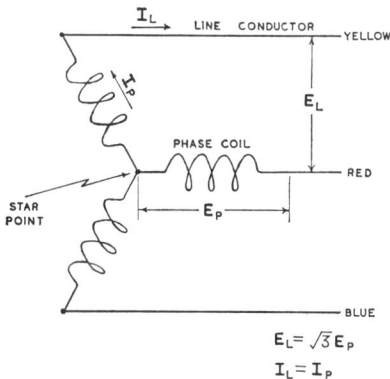

$$E_L = \sqrt{3}\, E_P$$
$$I_L = I_P$$

FIGURE 9.17   Three Phase Star Connection.

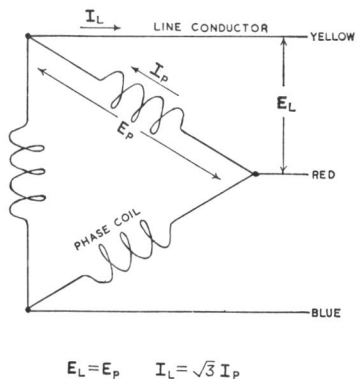

$$E_L = E_P \qquad I_L = \sqrt{3}\, I_P$$

FIGURE 9.18   Three Phase Delta Connection.

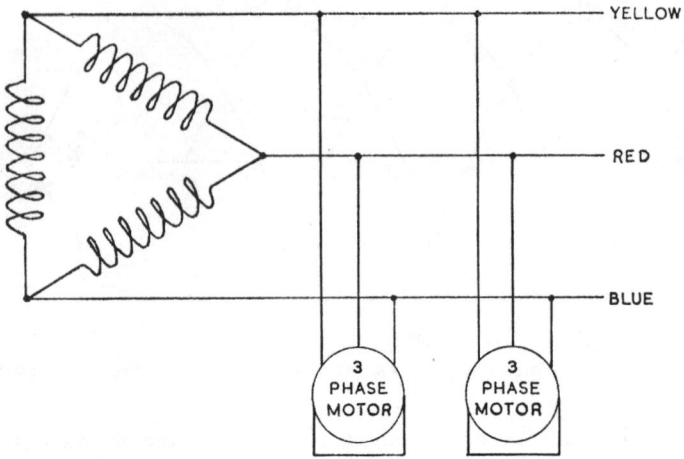

FIGURE 9.19   Three Phase Three-Wire Supply System.

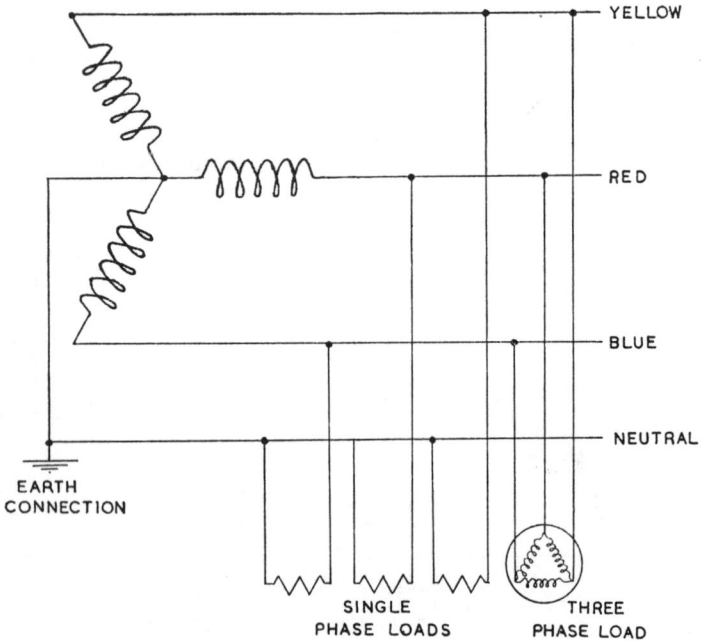

FIGURE 9.20   Three Phase Four-Wire Supply System.

the phase voltage and the line current is equal to $\sqrt{3}$ times the phase current.

$$E_L = E_P$$
$$I_L = \sqrt{3}\, I_P$$

9.33 A three-phase supply may be provided using either a *three-wire* or a *four-wire* system. A three-wire system is one in which only the three line conductors are used as illustrated in Fig. 9.19; the alternator (or transformer) providing the supply may use either star or delta connection. A four wire system is one in which the three line conductors and an extra conductor connected to the star point of the supply are used. This additional conductor is usually earthed at the supply end and is known as the *neutral* conductor. Figure 9.20 illustrates the four-wire system, and it can be seen that single phase loads can be connected between any line conductor and the neutral conductor, while three-phase loads can be connected to the three-line conductors. It should be noted that only star connected transformers or alternators may be used to supply a four-wire system as a delta connection would not provide a star point to which the neutral could be connected. The three-phase loads however can employ either star or delta connections as required.

Example 9.5

The pressure between the lines of a three-phase four-wire supply system is maintained at 415 V. Show how (a) a single phase lighting load (b) a three phase motor would be connected to the system and in each case state the value of electrical pressure supplied to the load.

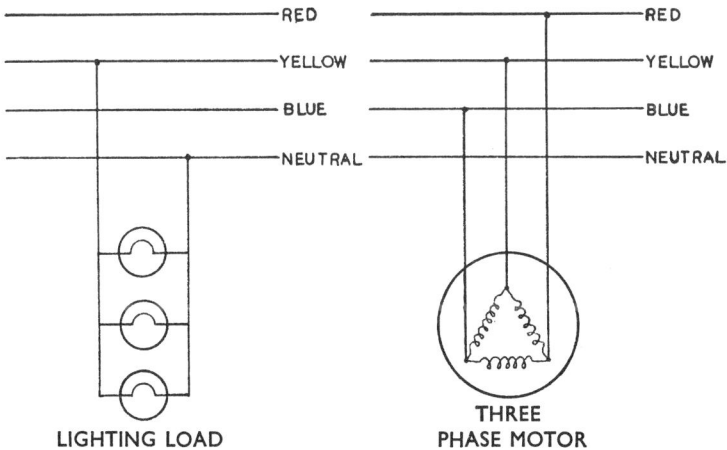

FIGURE 9.21   Example 9.5.

Solution

(a) The lighting load is connected between any phase conductor and the neutral conductor and so receives phase voltage.

$$E_P = \frac{E_L}{\sqrt{3}}$$

$$E_P = \frac{415}{\sqrt{3}}$$

$$\underline{E_P = 240 \text{ V}}$$

(b) The motor is connected to the three-line conductors and so receives line voltage.

$$\underline{E_L = 415 \text{ V}}$$

9.34 When four-wire connections are used, the neutral conductor may be regarded as providing a common return path for the currents in the three line conductors. It follows that the current in the neutral conductor must be the phasor sum of the currents flowing in the three line conductors. This phasor sum can be obtained by scaled drawing as follows. The three line currents are represented to scale and at their correct phase angles; then any two line currents are added, using the parallelogram method, to form a partial resultant. The partial resultant is in turn added to the third line current, again using the parallelogram method, so obtaining the final resultant which is the neutral current. The method is illustrated by Example 9.6 for a situation in which a different current flows in each line. It is left as an exercise for the student to show that if there are three equal currents at equal phase angles (i.e. a balanced load) then the partial resultant formed by the phasor addition of any two of the currents is always equal in magnitude but 180° out of phase with the third line current, so that the final resultant is zero. The neutral current for a balanced load is *zero*.

Example 9.6

In a 415 V three-phase four-wire system, non-inductive loads are connected between the line wires and the neutral such that the currents in the line wires are respectively 50 A, 30 A and 20 A. Find, by calculation or graphically, the value of the current in the neutral.　　　C.G.L.I.

Solution

Figure 9.22(a) shows the graphical solution; it is assumed that the current values have been given in the order corresponding to standard phase rotation: R, Y, B. Note that as the loads are non-inductive the
210

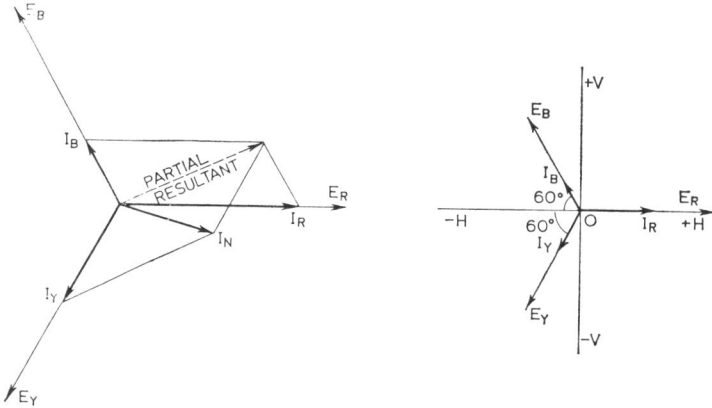

Figure 9.22   Example 9.6

phase angles are 0° and so each current phasor is directed along its appropriate phase voltage phasor. The diagram is started by drawing the three voltage phasors at a mutual angle of 120°. (Note that "B" is the "leading" phase). The currents are then added. (Note that phase angles are measured from the appropriate phase voltage, clockwise = lagging. In this case all phase angles = 0°.)

A parallelogram is formed using any two line currents, in this case I_R and I_B have been chosen; the partial resultant so obtained is then used to form a second parallelogram with the third current I_Y, and the final resultant is the neutral current, I_N. It can be seen that $I_N = 26.4$ A.

A solution can also be obtained by resolving the currents into components with respect to the horizontal and vertical axes shown in Fig. 9.22(b). The working is conveniently set out in tabular form as below, the + and − signs being obtained by inspection of the diagram.

Current	Horizontal Component	Vertical Component
I_R	$50 \cos 0 = 50$	$50 \sin 0 = 0$
I_Y	$-30 \cos 60 = -15$	$-30 \sin 60 = -25.98$
I_B	$-20 \cos 60 = -10$	$+20 \sin 60 = 17.32$
Total $= I_N$	25	-8.66

From above, using Pythagoras' theorem:

$$I_N = \sqrt{25^2 + (-8.66)^2}$$

$$I_N = \sqrt{700} = 26.4 \text{ A}$$

211

9.4 Three Phase Power

9.41 In general the power supplied by a three phase system is the sum of the powers supplied by each of the three phases. A special case which often arises is when the load is *balanced* that is when each phase supplies the same power at the same power factor. Most three phase motors automatically provide a balanced load. When a large number of single phase loads, such as lighting and heating apparatus, are connected to a three phase system it is good practice to arrange that each phase supplies approximately the same load. It can be shown that for a balanced three phase load, the total power consumed is given by the formula:

$$P = \sqrt{3}\, E_L I_L \cos \phi$$

Example 9.7

A three-phase induction motor is connected to a three-phase three-wire a.c. supply, the pressure between the lines being maintained at 415 V. The stator coils of the motor are connected in star for starting and in delta when the motor is running normally.

  (a) Draw circuit diagrams showing how the motor stator coils are connected (i) for starting, (ii) when running normally, and determine the voltage applied to each motor coil in each case.
  (b) If, when the motor is running on full load the current in each line is 12 A and the power factor is 0·8 lagging, determine the power consumed by the motor on full load.

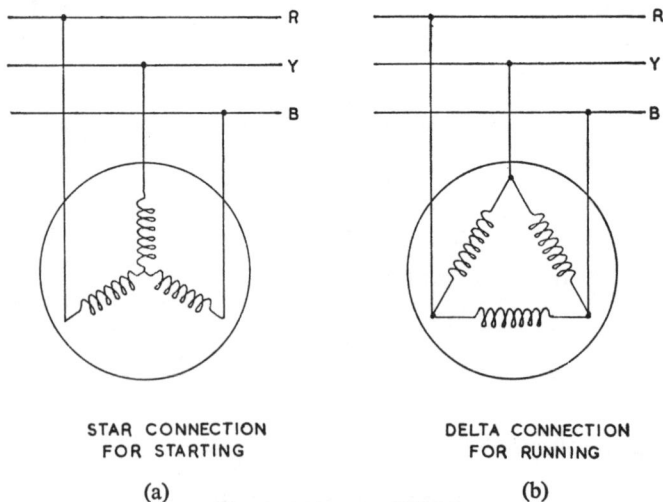

STAR CONNECTION   DELTA CONNECTION
FOR STARTING   FOR RUNNING

(a)   (b)

FIGURE 9.23 Example 9.7.
(a) Star connection for starting.
(b) Delta connection for running.

212

Solution

(a) Voltage across stator coils $= E_P$

When starting (star connection): $E_P = \dfrac{E_L}{\sqrt{3}}$

$$E_P = \dfrac{415}{\sqrt{3}}$$

$$\underline{E_P = 240 \text{ V}}$$

When running (delta connection): $E_P = E_L$

$$\underline{E_P = 415 \text{ V}}$$

(b)

$$P = \sqrt{3}\, E_L I_L \cos \phi$$

$$P = \sqrt{3} \times 415 \times 12 \times 0{\cdot}8$$

$$\underline{P = 6910 \text{ W}}$$

SUMMARY OF FORMULAE

Components of current:

I active $= I \cos \phi$

I reactive $= I \sin \phi$

$I = \sqrt{I^2 \text{ active} + I^2 \text{ reactive}}$

Components of voltage:

V active $= V \cos \phi$

V reactive $= V \sin \phi$

$V = \sqrt{V^2 \text{ active} + V^2 \text{ reactive}}$

Components of impedance:

$R = Z \cos \phi$

$X = Z \sin \phi$

$Z = \sqrt{R^2 + X^2}$

Components of power:

$P = \text{V A} \times \cos \phi$

$\text{V Ar} = \text{V A} \times \sin \phi$

$\text{V A} = \sqrt{P^2 + \text{V Ar}^2}$

Three phase star connection:

$E_L = \sqrt{3}\, E_P$

$I_L = I_P$

Three phase delta connection: $\qquad E_L = E_P$

$$I_L = \sqrt{3}\, I_P$$

Three phase power: $\qquad P = \sqrt{3}\, E_L I_L \cos\phi$
(for a balanced load.)

EXERCISE 9

1. The lighting load current of a workshop is 15 A in phase with the voltage and the power load current is 10 A lagging the voltage by 40°. Draw the phasor diagram of the currents to a scale of 1 mm = 0·2 A, and hence find the total current taken from the supply and the phase angle of this current relative to the supply voltage.

2. (a) Define *frequency*, *phase* and *cycle* as applied to an alternating quantity.
   (b) An a.c. motor takes a current of 10 A which lags the supply voltage by 45°. A capacitor connected in parallel with the motor takes a current of 6 A leading the supply voltage by 90°. Draw a phasor diagram using the supply voltage as reference phasor, and from it determine the total current consumed and its phase angle relative to the supply voltage.

3. Explain, with the aid of a phasor diagram, the meaning of power factor in the alternating current circuit. Why is a low power factor undesirable?
   A single phase load of 20 kW, at a power factor of 0·72, is supplied at 240 volts a.c. Calculate the decrease in current if the power factor is changed to 0·95, with the same kW loading. C.G.L.I.

4. Explain clearly the meaning of *power-factor* and illustrate by means of a phasor diagram.
   At full load, and at 0·766 power-factor lagging, the efficiency of a 30 kW output, 240 V single-phase motor is 87 per cent. Calculate the value of the supply current at full load:
   (a) with lagging power-factor as above,
   (b) when power factor correction is applied at the terminals of the motor so that the voltage and current are in phase,
   (c) when the power-factor is over-corrected so that the current leads the voltage by 40°. C.G.L.I.

5. The current taken by a 240 V, 50 Hz, single-phase induction motor running at full load is 39 amperes at 0·75 power-factor lagging. Calculate the intake taken from the supply (a) in kW, (b) in kVA.
   Find what size condenser connected across the motor terminals would cause the intake in kVA to be equal to the power in kW. C.G.L.I.

6. A single phase a.c. motor, operated from a 240 V, 50 Hz a.c. supply consumes 300 W at 0·5 p.f. lagging. Calculate the current consumed by the motor.
   If a 16 μF capacitor were connected across the motor terminals what would be the new values of the supply current and the overall power factor?

7. The following loads are installed in a workshop:
   (a) Lighting load, 5 kW, unity power factor
   (b) Power load, 7 kW at 0·6 p.f. lagging.
   Determine the total loading in kW and kVA and the overall power factor.

8. A group of single-phase motors connected to a 240 V, 50 Hz supply, consumes a total current of 30 A at 0·7 power-factor lagging, when all the motors are in use. What value capacitor could be used to correct the overall power-factor to 0·96 lagging with all the motors in use?

   If the motor loading were reduced so that without the capacitor in circuit they consumed 20 A at 0·5 power-factor lagging, what would be the overall current consumed with the capacitor connected?

9. A group of single-phase motors takes 50 A at 0·5 power factor lagging from a 240 V 50 Hz supply. Determine the total power consumed in kW and kVA.

   What size capacitor is required to improve the overall power factor to 0·92 lagging?

10. (a) What is the meaning of "power-factor"?
    (b) The full load efficiency of a 30 kW output, 415 V, 3-phase motor is 85 per cent. The current to the motor lags by 30 degrees.
        (i) Calculate the value of the full-load current.
        (ii) What would be the value of the current if the voltage and current were in phase.
        (iii) The current were leading by 30 degrees?   C.G.L.I.

11. What is meant by the terms *three-wire* and *four-wire* system as applied to a three-phase supply arrangement.

    Show by means of a diagram how a single-phase lighting and a three-phase motor load can be connected to a three-phase four wire supply, and determine the voltage available in each case if the supply voltage is 380 V between lines.

12. A three-phase motor consumes 20 kW at 0·8 power factor lagging from a 415 V three-wire three-phase supply. Calculate the kVA and the line current consumed by the motor. What is the current in each phase coil of the motor assuming that the coils are delta connected?

215

10 Alternating Current-Plant

10.1 Transformer Fundamentals

10.11 A *transformer* is an item of alternating current plant which takes in electrical energy at one pressure and gives the energy out again at some other pressure. A transformer has no moving parts and so is highly efficient, thus the power output from a transformer is very nearly equal to the power input. The use of transformers makes it possible to transmit electrical energy at a high pressure, the supply pressure being eventually reduced by the use of transformers to a level suitable for the consumer. The advantage of high voltage transmission is that less current is required to produce a certain power at a high pressure than at a lower pressure. By operating transmission lines at a high pressure, the current flowing in them is reduced, leading to a reduction both in the size of cable required and in the voltage drop and energy losses in the system.

10.12 The double wound transformer consists essentially of two coils, insulated from each other, and mounted on a common magnetic core as shown in Fig. 10.1. When an alternating e.m.f. is applied to the input or *primary* winding, the current, which then flows through this winding, produces an alternating magnetic flux in the iron core. The magnetic flux

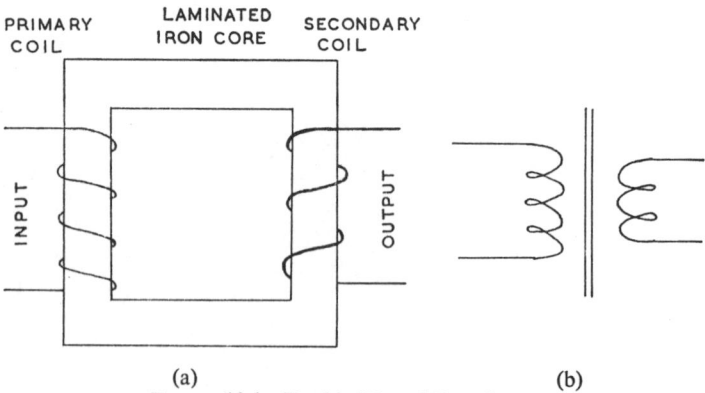

(a)   (b)

FIGURE 10.1   Double Wound Transformer.

(a) Simple Transformer.
(b) Theoretical Symbol.

in turn induces a back e.m.f. in the primary coil which, by acting in opposition to the supply e.m.f., limits the current in the primary winding. This effect is exactly the same as that of self-inductance (see Section 7.5), and in a perfect transformer the induced e.m.f. would be equal in value to the e.m.f. applied to its terminals. At the same time the magnetic flux induces an e.m.f. in the secondary winding, so that there is mutual inductance between the coils. (See Section 7.5.) The e.m.f. induced in a coil depends on the number of turns in the coil and the rate of change of the magnetic flux linked with them. As both coils of the transformer are linked by the same flux, their induced e.m.fs. will be proportional to the number of turns in each coil and so:

$$\frac{E_p \text{ (Primary e.m.f.)}}{E_s \text{ (Secondary e.m.f.)}} = \frac{N_p \text{ (Primary turns)}}{N_s \text{ (Secondary turns)}}$$

Example 10.1

A transformer connected to a 240 V, 50 Hz a.c. supply has a primary winding of 1200 turns.

(a) What e.m.f. is induced in a secondary winding of 300 turns?
(b) How many turns would be required in the secondary winding if it is to produce an e.m.f. of 100 V?

Solution

(a)
$$\frac{E_p}{E_s} = \frac{N_p}{N_s}$$
$$\frac{240}{E_s} = \frac{1200}{300}$$
$$E_s = \frac{240 \times 300}{1200}$$
$$E_s = 60 \text{ V}$$

(b)
$$\frac{E_p}{E_s} = \frac{N_p}{N_s}$$
$$\frac{240}{100} = \frac{1200}{N_s}$$
$$N_s = \frac{1200 \times 100}{240}$$
$$N_s = 500 \text{ turns.}$$

In many practical transformers, connections called tappings are provided on the coils. In this way it is possible to utilise a portion of the total number of turns available and so alter the voltage ratio of the transformer.

Example 10.2

A bell transformer is connected to a 240 V, 50 Hz a.c. supply. The primary winding has 1800 turns and the secondary winding provides an output at 12 V with a tapping at 4 V. Determine the number of turns in the secondary winding and the position of the tapping point. With the aid of a diagram show how connections may be made to obtain each of the secondary voltages.

Solution

$$\frac{E_p}{E_s} = \frac{N_p}{N_s}$$

$$\frac{240}{12} = \frac{1800}{N_s}$$

$$N_s = \frac{1800 \times 12}{240}$$

$$N_s = 90 \text{ turns}$$

In order to find the position of the tapping point, it is necessary to calculate the number of turns needed to give 4 V, this gives:

$$N_s = \frac{1800 \times 4}{240}$$

$$N_s = 30 \text{ turns}$$

i.e. the tapping point must be 30 turns from one end of the winding.
Figure 10.2 shows how the different secondary voltages may be obtained.

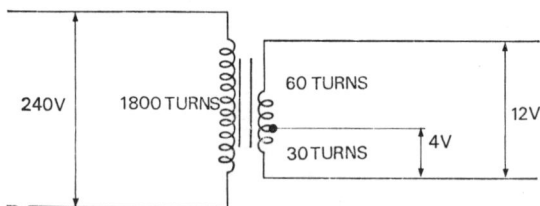

FIGURE 10.2 Example 10.2.

10.13 When a current is taken from the secondary winding, the magnetizing force caused by the flow of this current tends to reduce the magnetic flux in the core. This in turn tends to reduce the e.m.f. induced in the primary winding, so allowing more current to flow into the winding from the supply. The increase in primary current restores the magnetic flux to its original value, so that the e.m.fs. of both windings remain unchanged. Thus the de-magnetizing effect of the secondary current is balanced by the magnetizing effect of the primary current.

$$\text{Primary ampere turns} = \text{Secondary ampere turns}$$
$$I_p N_p = I_s N_s$$
$$\frac{I_p}{I_s} = \frac{N_s}{N_p}$$

Example 10.3

Make a neat sketch or diagram of a single phase double-wound transformer, and describe its action, clearly explaining the function of each part.

The "volts per turn" of a certain single phase transformer are 1·7. The transformer has a step down ratio of 3825 V to 255 V. Calculate: (a) the respective number of turns in each winding, and (b) the secondary current if the primary current is 12 A.   C.G.L.I.

Solution

The three main parts of a double wound transformer are the primary coil, the secondary coil and the laminated iron core. When an a.c. supply of e.m.f. E_p is connected to the primary coil, an alternating magnetic flux Φ is set up in the laminated iron core. This magnetic flux in turn induces an e.m.f. E_s in the secondary coil, which is used to supply the load. Thus the primary coil is connected to the supply and its purpose is to magnetize the core. The core provides a good magnetic circuit linking the two coils. The secondary coil is responsible for providing the output by virtue of the e.m.f. induced in it. It can be shown that for a perfect transformer:

$$\frac{E_p}{E_s} = \frac{N_p}{N_s}$$

and

$$\frac{I_p}{I_s} = \frac{N_s}{N_p}$$

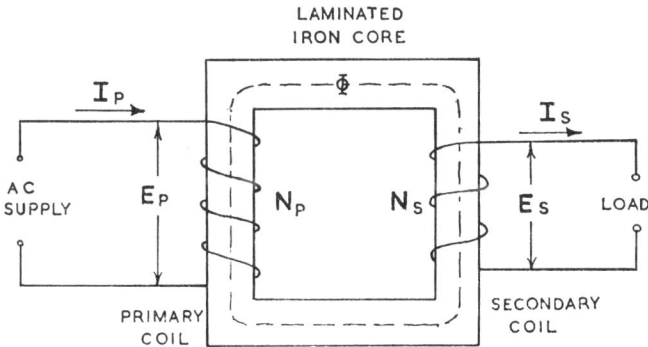

FIGURE 10.3   Example 10.3. Single Phase Double-Wound Transformer.

219

(a) As each turn has an induced e.m.f. of 1·7 V then:

$$\text{Primary turns } (N_p) = \frac{3825}{1·7}$$

$$\text{i.e.} \quad \underline{N_p = 2250 \text{ turns}}$$

$$\text{Secondary turns } (N_s) = \frac{255}{1·7}$$

$$\text{i.e.} \quad \underline{N_s = 150 \text{ turns}}$$

(b)
$$\frac{I_p}{I_s} = \frac{N_s}{N_p}$$

$$\therefore \quad I_s = \frac{I_p N_p}{N_s}$$

$$\text{i.e.} \quad I_s = \frac{12 \times 2250}{150}$$

$$\text{Hence: } \underline{I_s = 180 \text{ A}}$$

10.14 In the ordinary double-wound transformer there is no direct electrical connection between the two windings but, in the auto-transformer, there is only one winding which has one or more tapping points. This winding serves as both primary and secondary, the basic connections for step-up and step-down auto-transformers being shown in Fig. 10.4. The auto-transformer is less expensive than a double-wound transformer but its use is limited because of the dangers inherent in the direct electrical connections which exist between the input and output terminals. Auto-transformers are often employed in a.c. motor starters to reduce the

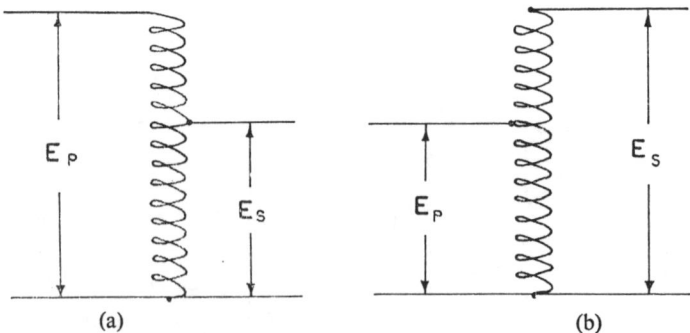

FIGURE 10.4 Auto-transformer Connections.
(a) Step-down Auto-transformer.
(b) Step-up Auto-transformer.

pressure applied to the motor during the starting period. They are also useful as booster transformers to give a small step-up to the pressure in a mains supply system.

Example 10.4

An auto-transformer has a coil of 500 turns, tapped at 480 and 440 turns. A 240 V a.c. supply is connected between the common terminal and the tapping at 480 turns. What e.m.f. is induced between the common terminal and (a) the tapping at 440 turns, (b) the tapping at 500 turns.

Solution

(a) $E_s = E_p \times \dfrac{N_s}{N_p}$

$E_s = 240 \times \dfrac{440}{480}$

$\underline{E_s = 220 \text{ V}}$

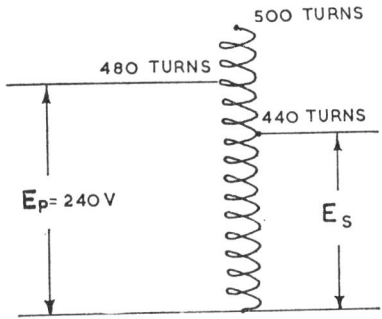

FIGURE 10.5 Example 10.4 (a).

(b) $E_s = E_p \times \dfrac{N_s}{N_p}$

$E_s = 240 \times \dfrac{500}{480}$

$\underline{E_s = 250 \text{ V}}$

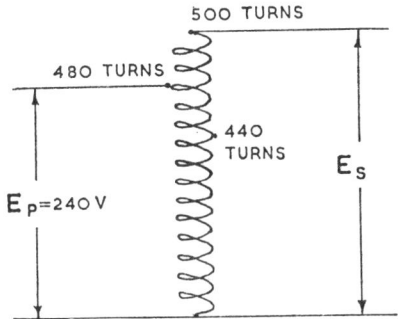

FIGURE 10.6 Example 10.4 (b).

10.2 Transformer Construction

10.21 The cores of power transformers are usually constructed using silicon steel, as this material has a high magnetic permeability and a low hysteresis loss when carrying alternating magnetic flux. If the core were solid a current would be induced in it circulating in the same direction as the current in the secondary coils as shown in Fig. 10.7(a). This circulating or eddy current would heat the core and cause a substantial power loss.

221

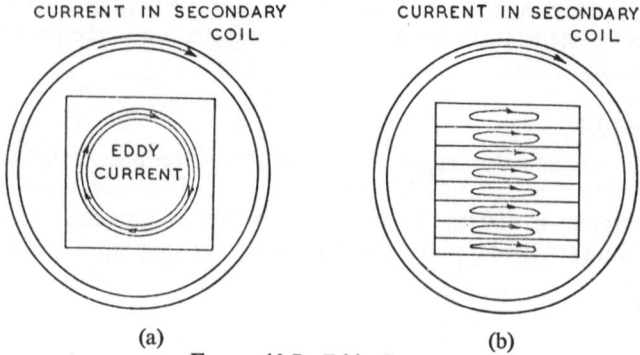

FIGURE 10.7   Eddy Currents.
(a) Solid Core Provides Easy Path for Eddy Current.
(b) Laminated Core Forces Eddy Currents into Long Narrow Paths.

In order to reduce the eddy current loss to the lowest practical value, the cores of transformers are laminated as shown in Fig. 10.7(b) so forcing the eddy currents into long narrow paths and reducing their value considerably. There are two main shapes of core used for single phase transformers. The *shell type* core has three limbs, the centre limb being of a greater cross-sectional area than the outer limbs. Both primary and secondary coils are wound over the centre limb, no coils being wound on the outer limbs. The *core type* core has only two limbs, primary and secondary coils being wound on each limb. These two types of core are illustrated by Fig. 10.8(a) to (d). Small transformers are sometimes wound on ring type cores, made by winding a length of silicon steel strip into the form of a ring.

10.22   The windings of practical transformers are distributed equally over all the limbs. Thus in a core type double-wound single phase transformer, which has two limbs, half of the primary and half of the secondary turns are wound on one limb, the remainder of the turns being wound on the other limb. The coils may be wound concentrically, one winding on top of the other as in Fig. 10.9(a), or side by side as in Fig. 10.9(b). The windings are of copper wire or strip, insulated by cotton, paper, varnished cambric or simply varnish in the case of small transformers. Additional insulation is provided between the coils using press board or synthetic resin bonded cylinders in large transformers although, in small sizes, varnished cambric is sometimes used for this purpose.

10.23   In every transformer some power loss occurs, the principal types of loss being:

(a) the *iron loss* which is due to eddy currents and hysteresis in the core and is constant at all loads, and
(b) the *copper loss* which is due to the resistance of the windings and is proportional to the square of the load current.

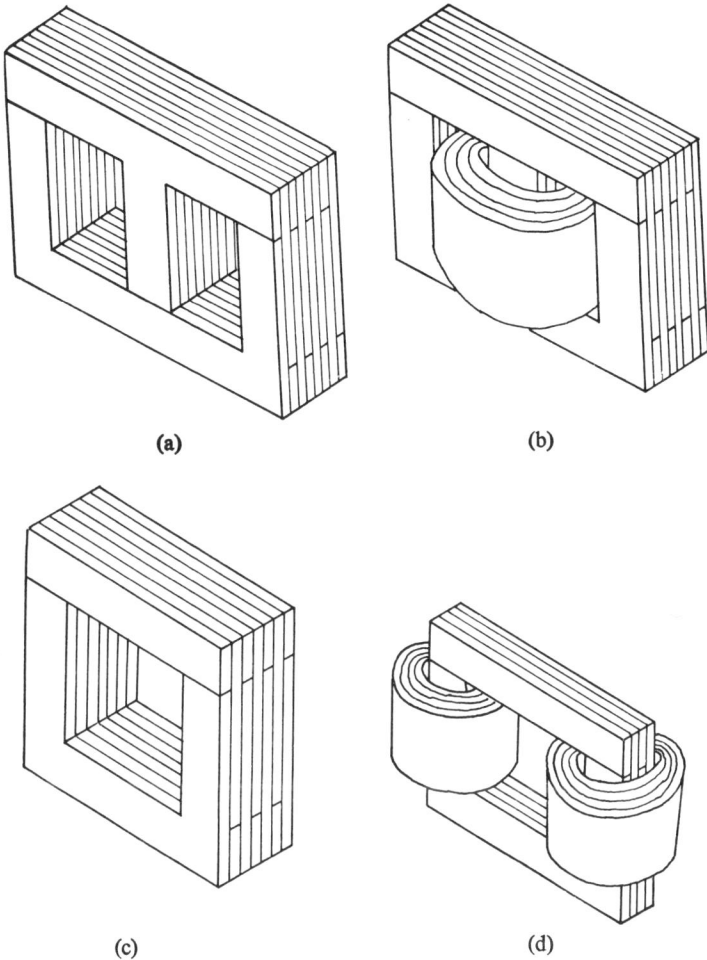

FIGURE 10.8   Transformer Cores.
(a) Shell Type.   (b) Shell Type with Coils Fitted.
(c) Core Type.   (d) Core Type with Coils Fitted.

The effect of these losses means that the efficiency of a transformer, although usually very high, can never be 100%. Also the power loss is converted into heat and so tends to raise the temperature of the transformer. Small transformers may well have sufficient natural cooling because of the air flow past them, but special cooling arrangements have to be made for larger transformers. One common method is to immerse the transformer in an oil tank, for the oil not only provides a cooling medium but also acts as an insulator. For the larger transformers the cooling provided by the

223

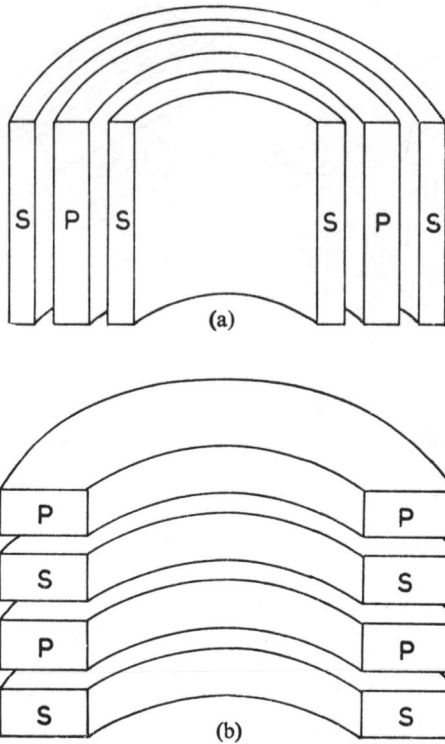

FIGURE 10.9   Transformer Windings.
(a) Concentric Winding.   (b) Sandwich Winding.

surface area of a plain tank is insufficient and cooling tubes are often fitted, as shown in Fig. 10.10.

10.24   When a transformer supplies a load current there is a voltage drop in its windings. This voltage drop is caused partly by the resistance of the windings and partly by an effect called *leakage inductance*, which is due to the magnetic flux in the core not perfectly linking with the coils. The voltage drop is usually small even at full load and is often expressed as a fraction of the no-load voltage. The fractional voltage drop is called the *regulation* of the transformer and so:

$$\text{Regulation} = \frac{\text{Volts on no-load} - \text{Volts on load}}{\text{Volts on no-load}}$$

In many cases the regulation of a transformer is determined at full load and is expressed as a percentage, which is obtained by multiplying the full-load fractional value by 100.

224

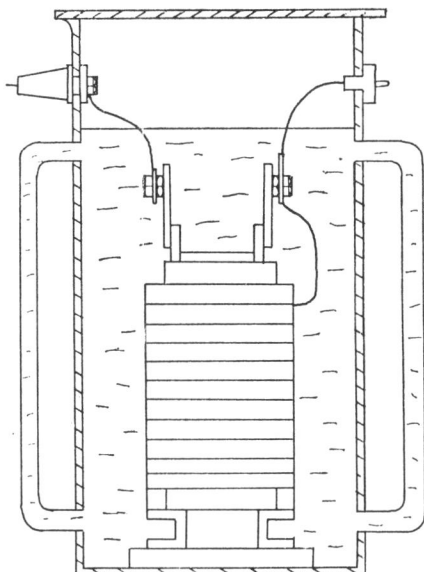

FIGURE 10.10   Oil Cooled Transformer.

Example 10.5

A transformer provides an output of 250 V on no-load and 240 V on full-load. What is its percentage regulation at full load?

Solution

$$\text{Full load regulation} = \frac{\text{Volts on no-load} - \text{Volts on full-load}}{\text{Volts on no-load}}$$

$$\text{Full load regulation} = \frac{250 - 240}{250}$$

Full load regulation $= 0.04$
Converting this to a percentage:

Full load percentage regulation $= 0.04 \times 100$

Full load percentage regulation $= 4\%$

10.3   Rotating Magnetic Fields

10.31   Many types of a.c. motor make use of the fact that it is possible to use an a.c. supply to produce a rotating magnetic field. If three coils are spaced at 120° intervals as shown in Fig. 10.11(a) and each coil is

225

(a)

(b)

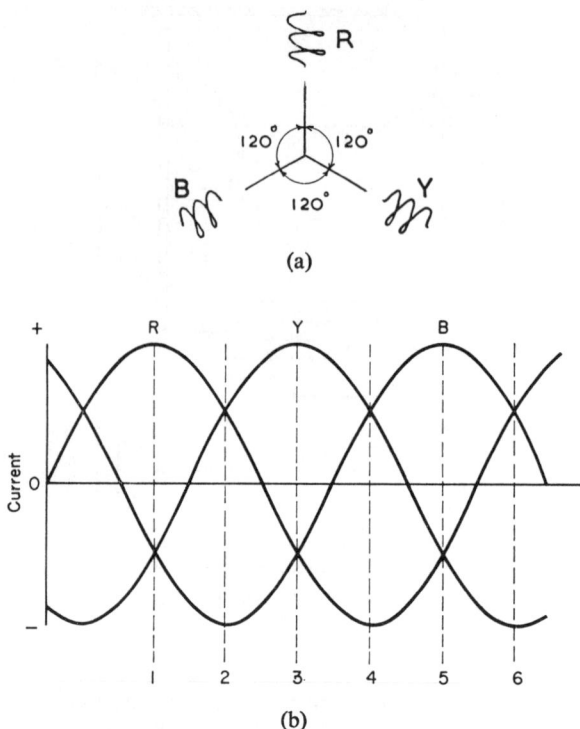

FIGURE 10.11   Production of Rotating Magnetic Field.
    (a) Arrangement of Coils.
    (b) Wave Diagram of Three Phase Currents.

connected to one phase of a three phase supply, then the currents flowing through the coils vary from instant to instant as shown by the wave diagram of Fig. 10.11(b).

10.32   Consider the time-instant marked 1 on the wave diagram. The red phase coil is carrying maximum current in the positive direction so producing a peak magnetizing force in line with itself. At the same time-instant the blue and yellow phase coils are each carrying a negative current of half maximum value, so producing a half-peak value magnetizing force in line with each coil. Figure 10.12(a) shows the three coils and the directions of the magnetizing forces at time-instant 1. These forces can be added using vectors as shown in Fig. 10.12(b), where OA represents the magnetizing force of the red phase coil in magnitude and direction. Similarly AB represents the magnetizing force of the blue phase coil and BC represents the magnetizing force of the yellow phase coil. OC represents the resultant magnetizing force of the three coils acting together.

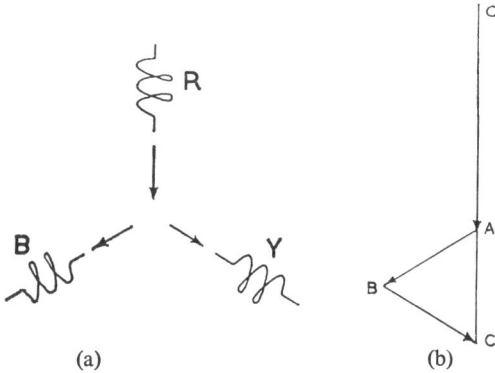

[FIGURE 10.12 Magnetizing Force Vectors.
(a) Directions of Magnetizing Forces.
(b) Vector Addition of Magnetizing Forces.

10.33 By repeating the process the resultant magnetizing force for other instants of time can be found. Figure 10.13 shows the direction of the resultant magnetizing force for the time instants marked 1 to 6 on the wave diagram. From the above it can be seen that a three-phase winding can produce a magnetic field which remains constant in strength but whose direction rotates at a constant speed. In practice a rotating magnetic field can be produced by a three-phase winding situated in slots in the inner circumference of the fixed part of the motor which is called the stator. The field produced may be a two-pole field, making one revolution per cycle, a four-pole field making one revolution in two cycles, and so on. Any even number of poles may be used, a "p"-poled field making one revolution in $p/2$ cycles. Thus the speed at which the field rotates is $2f/p$ revolutions per second, where f is the supply frequency,

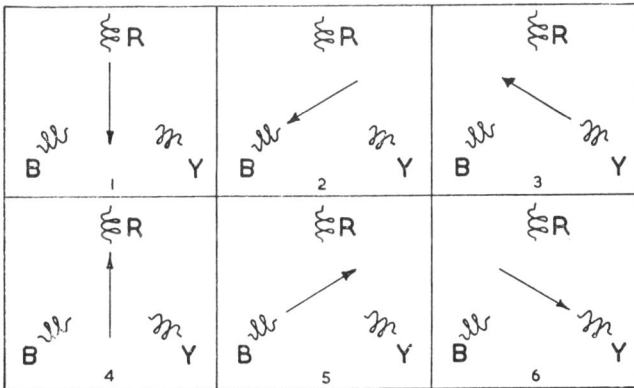

FIGURE 10.13   Resultant Magnetizing Forces at Various Time Instants.

This speed is known as the synchronous speed and is more conveniently calculated from the formula:

$$N_s = \frac{60 \times 2f}{p} \text{ rev/min}$$

Two types of a.c. motor which use a rotating magnetic field are the synchronous motor and the induction motor.

10.4 A.C. Motors

10.41 The *synchronous motor* has a rotating part or "rotor" which is magnetized with a fixed polarity. In small motors the rotor may be permanently magnetized while in larger motors the rotor carries a winding which is fed with direct current via two slip rings. This type of motor will run at synchronous speed only when the permanent field of the rotor "locks" with the rotating magnetic field of the stator. Small synchronous motors are used in electric clocks because their speed is completely governed by the frequency of the main supply. Larger types of synchronous motor are sometimes used because they may be made to operate at a leading power factor, but their use is limited owing to the complicated starting equipment required.

10.42 The *induction motor* is the most widely used type of a.c. motor owing to its comparatively low cost and high efficiency. Two types of rotor may be used:

> (a) The "squirrel cage" rotor.
>
> (b) The "wound" rotor.

The winding of a squirrel cage rotor usually consists of copper or aluminium bars fitted in slots, the bars being connected to end rings at each end of the rotor. With this type of rotor the rotor bars are not connected directly to the source of supply so that the operation of the motor depends on currents induced in the rotor bars by the rotating magnetic field. In order to induce the necessary rotor currents, the magnetic field must rotate faster than the rotor, therefore the actual speed (N_r) of an induction motor is always slightly below the synchronous speed (N_s). The fractional slip of an induction motor is defined as:

$$\text{Fractional slip} = \frac{N_s - N_r}{N_s}$$

Example 10.6

A four pole, three-phase induction motor is connected to a 415 V, 50 Hz, three-phase supply, calculate:

(a) the percentage slip when the motor is running at 1450 rev/min.

(b) the motor speed when the slip is 6%.

Solution

First calculate the synchronous speed of the motor, which is:

$$N_s = \frac{60 \times 2f}{p}$$

$$N_s = \frac{60 \times 2 \times 50}{4}$$

$$N_s = 1500 \text{ rev/min}$$

Then:

(a)   Fractional slip $= \dfrac{N_s - N_r}{N_s}$

$$s = \frac{1500 - 1450}{1500}$$

$$s = 0 \cdot 0333$$

Percentage slip $= 0 \cdot 0333 \times 100$

Percentage slip $= 3 \cdot 33$

(b) $N_r = N_s - sN_s$

$$N_r = 1500 - \frac{6}{100} \times 1500$$

$$N_r = 1500 - 90$$

$$N_r = 1410 \text{ rev/min}$$

10.43 The squirrel cage induction motor is widely used as it is both cheap and efficient, having only one real disadvantage in that its starting

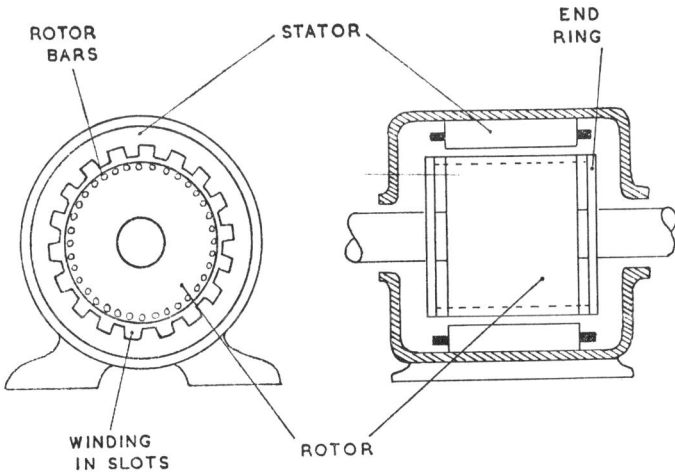

FIGURE 10.14   Induction Motor.

torque is comparatively low. When a high starting torque is needed then a slip ring motor, using a wound rotor, may be used.

The wound rotor has a winding made up of coils situated in slots. The coils are usually connected so that they form a three-phase star-connected winding with the outer ends of the coils brought out to three slip rings. When the motor is running at its normal speed the slip rings are short-circuited by the motor starter, and the motor operates in the same way as the squirrel cage motor. But, when starting, resistors are connected in series with the slip rings. This not only reduces the starting current but also increases the starting torque.

10.44 Small series motors will operate satisfactorily when connected to an a.c. supply. As the field windings and armature are connected in series they must carry the same current at all times and so, when the current in the field reverses, so also does the current in the armature, and the motor continues to run in the same direction. Motors intended to be used in this way are often called *universal motors* because they will operate either from an a.c. or a d.c. supply. It should be noted that a shunt connected motor cannot be operated satisfactorily from an a.c. supply. This is because the high inductance of the shunt field coils causes the current in them to lag the supply e.m.f. so that the currents in the armature and field no longer reverse at the same time. The field system of a universal motor is always built up of laminations in order to reduce the eddy currents which occur when the motor is operated from an a.c. supply.

10.45 A single-phase a.c. supply produces a pulsating rather than a rotating magnetic field, nevertheless single-phase a.c. induction motors will operate successfully provided that an initial start is given to the rotor. The single-phase motor will run equally well in either direction, so that the direction in which it rotates is determined by the direction of the initial starting torque. Two common methods of providing the necessary starting torque are:

(a) Split-phase start.

(b) Capacitor start.

Motors intended for "split-phase" starting have a "running" winding and a "starting" winding; the starting winding possesses a high inductance so that the current flowing through it lags the supply voltage by a large phase angle. This has an effect equivalent to a two-phase supply and so produces a rotating magnetic field which starts the motor. When the motor has gained sufficient speed a centrifugal switch disconnects the starting winding. Figure 10.15 shows the connections for a split-phase start motor. Capacitor start motors also have two windings, but in this case the starting winding has a comparatively low inductance and is connected in series with a capacitor. The series capacitor causes the current flowing

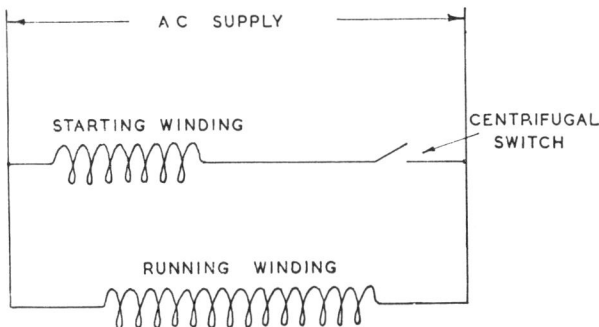

FIGURE 10.15   Split-Phase Start Motor.

through the starting winding to lead the supply voltage so, once again, the effect of a two-phase supply is obtained, resulting in the production of a rotating magnetic field which starts the motor. When the motor is up to speed, the starting winding is usually disconnected automatically by

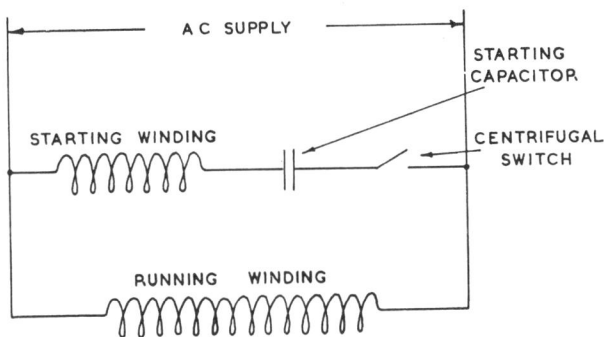

FIGURE 10.16   Capacitor Start Motor.

a centrifugal switch; but in some cases the starting winding may continue to receive current via a smaller capacitor. Figure 10.16 shows the connections of a capacitor start motor.

10.5   Rectification

10.51   The function of a rectifier is to convert an a.c. supply into a d.c. supply. Rectifier circuits employ devices which will allow current to pass in one direction only, the simplest rectifier circuit consisting simply of a rectifying device in series with the load, as shown in Fig. 10.17(b). It can be seen from the wave diagram of the rectified current, Fig. 10.17(d), that only the positive part of the a.c. input produces current in the load, the

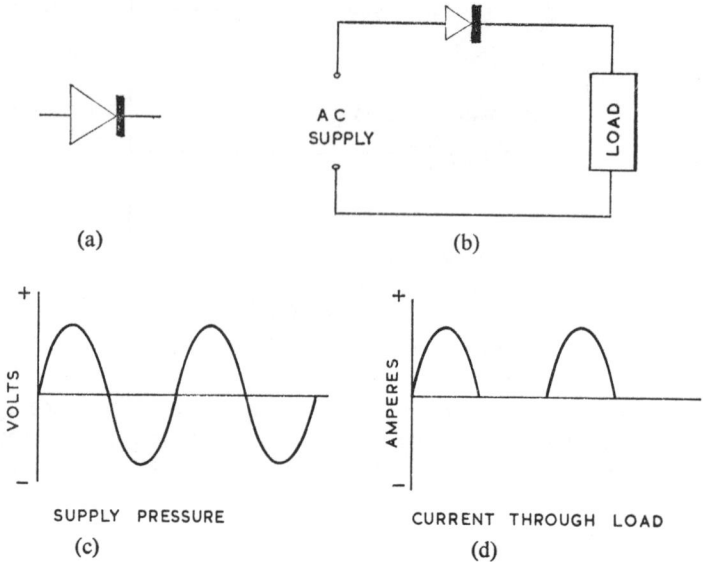

FIGURE 10.17   Half Wave Rectification.
(a) Symbol for Rectifying Device.
(b) Simple Rectifier Circuit.
(c) Waveform of Supply e.m.f.
(d) Waveform of Rectified Current.

flow of current being prevented during the negative part of the wave by the action of the rectifier. A circuit which operates in this way is called a half-wave rectifier circuit because the d.c. output is obtained from only one half of the a.c. input wave. A practical rectifier element should have a low *forward resistance*. That is, it should not present much resistance to current flow in the forward direction, as any such resistance causes a power loss which heats the rectifier element and lowers the efficiency. It should also be able to withstand a high voltage in the reverse direction without passing undue current, as again this causes over-heating and loss of efficiency. The maximum reverse voltage which a rectifier can withstand without damage is called the *reverse breakdown voltage*.

10.52   The output of a half-wave rectifier circuit consists of pulses of unidirectional current separated by periods, one half cycle long, during which there is no output. The "bridge" circuit, illustrated in Fig. 10.18, provides one method of obtaining an output from each half cycle of the a.c. input. As such a circuit provides an output from each part of the a.c. input wave, it is called a *full-wave rectifier* circuit. For the positive half of the input waveform, current flows through rectifier elements 1 and 3 as shown by the solid arrows; the current cannot flow through elements 2 and 4 during this half cycle as they will not allow current to flow in this

232

direction. During the next half cycle the polarity of the a.c. supply reverses and now drives current through rectifier elements 2 and 4 as shown by the dashed arrows, while elements 1 and 3 block the flow of current through themselves. It can be seen that the current through the load is always in the same direction no matter which half cycle produces it.

10.53 One form of rectifier element which may be used in the circuits shown in Figs. 10.17 and 10.18 is the *metal rectifier*. The *copper oxide*

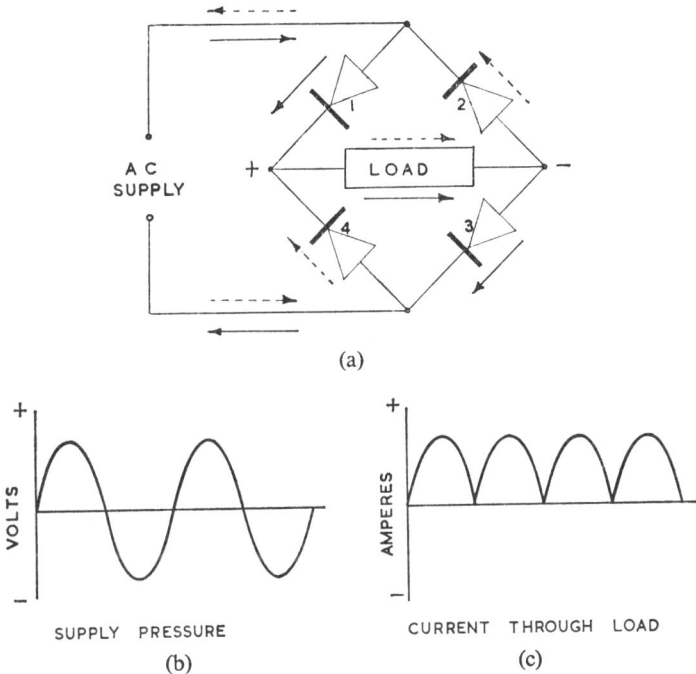

(a)

SUPPLY PRESSURE

CURRENT THROUGH LOAD

(b)   (c)

FIGURE 10.18   Bridge Rectifier Circuit.
(a) Circuit Diagram.
(b) Waveform of Supply e.m.f.
(c) Waveform of Load Current.

type of metal rectifier is illustrated in Fig. 10.19. Each rectifier element consists of a copper washer which has been oxidised on one side to form a layer of copper oxide, contact being made with the oxide layer by means of a lead washer. Current can flow easily through the element in the direction oxide to copper (forward direction) but cannot flow easily in the reverse direction. As a single copper oxide element can withstand a voltage in the reverse direction of only about 8 V, several elements are often mounted in series, as shown in Fig. 10.19(b), so that a higher voltage can be obtained. Cooling fins are often mounted between the rectifier elements as shown.

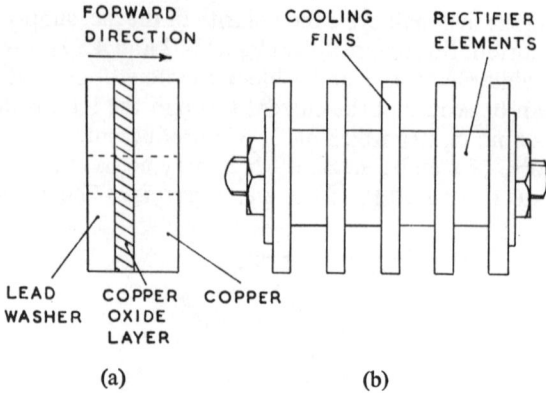

FIGURE 10.19   Copper Oxide Rectifier.
(a) Basic Rectifier Element.
(b) Several Elements Mounted in Series.

10.54 Another form of metal rectifier is the *selenium rectifier*. Each selenium rectifier element consists of an aluminium or nickle plated disc on which is deposited a thin layer of selenium. The selenium layer is then sprayed with a layer of low melting point alloy of tin and cadmium which forms the contact with the selenium layer. Current can flow easily in the direction selenium to alloy layer (forward direction), but cannot flow easily in the reverse direction. The permissible current through a selenium rectifier element is less than for a copper oxide rectifier of the same plate size, but the permissible reverse voltage is higher, about 30 V, so that fewer elements are needed in series to obtain the voltage rating required. Figure 10.20 illustrates the construction of the selenium rectifier.

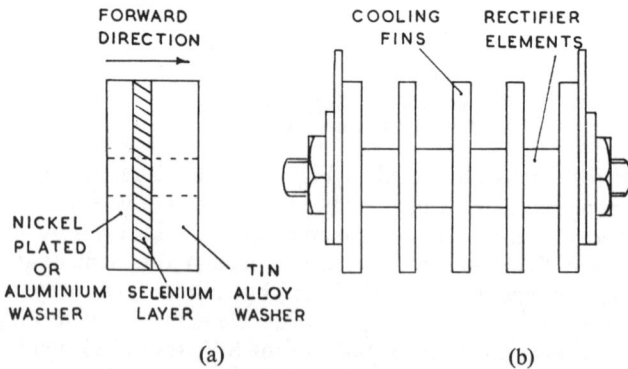

FIGURE 10.20   Selenium Rectifier.
(a) Basic Rectifier Element.
(b) Several Elements Mounted in Series.

10.55 In recent years *junction diode* rectifiers using semiconducting materials have been developed. One form consists of a thin slice of germanium crystal onto which a pellet of indium is fused to form the rectifying junction Alternatively a silicon crystal may be used, the junction being formed by fusing an aluminium pellet onto the crystal. Both silicon and germanium junction diodes are very small in size compared

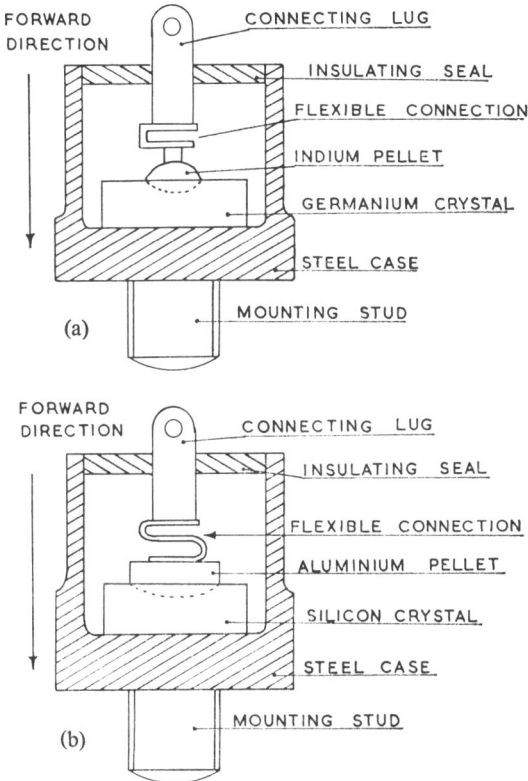

FIGURE 10.21   Junction Rectifiers.
(a) Germanium Junction Diode.
(b) Silicon Junction Diode.

with other types of metal rectifier of similar rating, the silicon junction diode having the advantage of a much higher permissible working temperature than the germanium type. Figure 10.21 shows the construction of germanium and silicon junction diodes. Semiconductor diodes can be designed to have a high reverse breakdown voltage so that it is not usually necessary to employ elements connected in series. Also the forward resistance can be lower than that of either copper oxide or

235

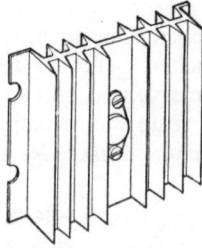

FIGURE 10.22 Heat Sink.

selenium types, leading to highly compact efficient rectifier installations. Owing to their small physical size, junction diode rectifiers often have insufficient surface area to dissipate the heat produced in them during normal operation without excessive temperature rise. For this reason these types of diode are often mounted on finned metal blocks called *heat sinks*. These effectively increase the area from which heat can be dissipated and so help to limit the temperature rise. Figure 10.22 shows a typical small heat sink.

10.6 Thermionic Diode Valve

10.61 The thermionic diode valve has two electrodes, the *cathode* and the *anode* mounted inside an evacuated glass envelope, as illustrated by Fig. 10.23. When the valve is in use the cathode is heated, and this causes

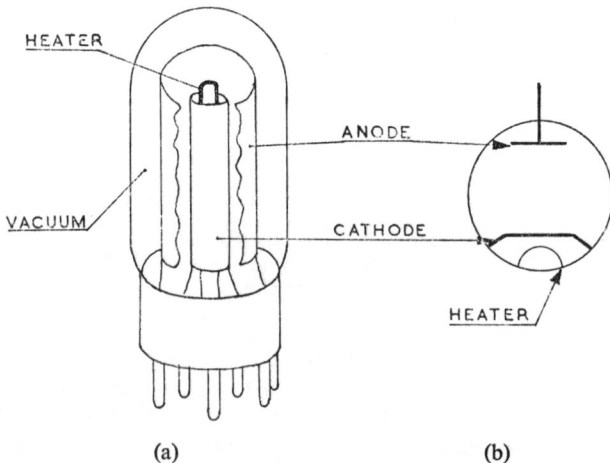

(a)   (b)

FIGURE 10.23 Thermionic Diode Valve.
(a) Construction.
(b) Symbol.

electrons to be emitted from it. The electrons form a *negative space charge* surrounding the cathode. If the anode is given a positive potential with respect to the cathode, it attracts the electrons (since they have a negative charge and unlike charges attract), so that a current flows through the valve. But if the anode is given a negative potential it repels the electrons (like charges repel) and no current can flow through the valve. Figure 10.24 illustrates the action of a thermionic diode valve. It can be seen that the electron flow through the valve can take place only in the direction cathode to anode. As electrons are negative electric charges this is equivalent to a conventional current flowing from anode to cathode, i.e., a negative current in one direction is equivalent to a positive current in the other direction.

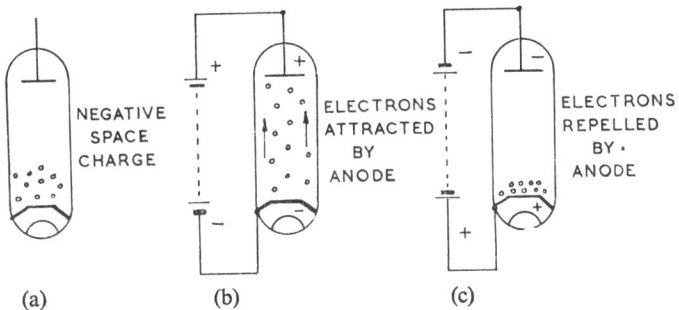

FIGURE 10.24 Action of Diode Valve.
(a) Electrons Forming Negative Space Charge when Cathode is Heated.
(b) Anode Made Positive so Causing Electron Flow.
(c) Anode Made Negative; No Electron Flow.

10.62 As the thermionic diode valve can pass current in one direction only, it can be used as a rectifier element. Figure 10.25 shows a simple half-wave circuit using a thermionic diode valve. It can be seen that the circuit is slightly more complicated than that shown in Fig. 10.17(b) owing to the necessity of supplying current to the cathode heater. In the circuit shown the cathode heater is supplied via a resistor R which is chosen to give the correct heater current for the valve. The action of the circuit is otherwise the same as for the circuit shown in Fig. 10.17.

10.63 Figure 10.26 shows a widely used circuit for obtaining full wave rectification using thermionic diode valves. This circuit uses two diodes which are often contained in one envelope, the complete valve being known as a double diode valve. As can be seen from the circuit diagram a transformer with a centre tapped secondary winding is used to supply the two anodes with voltages which are 180° out of phase with each other. Thus when a positive voltage is applied to the anode of one diode, so that this valve conducts, a negative voltage is applied to the anode of the other

237

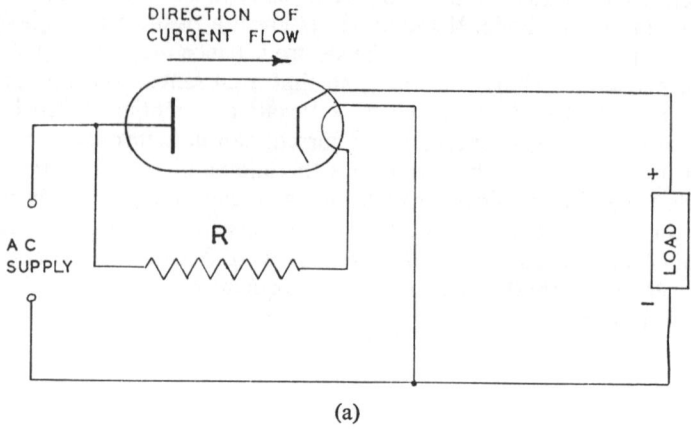

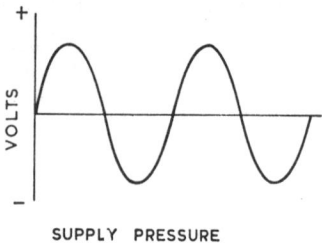

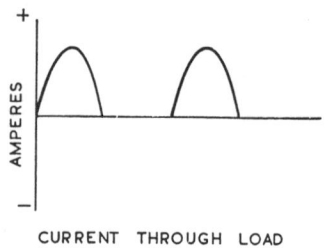

FIGURE 10.25   Half Wave Rectification by Thermionic Diode.
(a) Circuit Diagram.
(b) Waveform of Supply e.m.f.
(c) Waveform of Rectified Current.

diode so that this valve is not conducting. The two valves carry current alternately, the currents being combined to produce a full-wave rectified output. In the circuit shown the cathode heating circuit is supplied by a special winding on the transformer.

10.7 Smoothing

10.71 Although the current in a load supplied by a rectifier is uni-directional, that is it flows in one direction only, its value fluctuates so that it is not a true steady d.c. current. A *smoothing circuit* can be used to process the rectifier output and provide a load current which more nearly approaches an ideal steady d.c. current. Smoothing circuits can make use of the properties of either capacitance or inductance or a combination of both.

238

(a)

SUPPLY PRESSURE

(b)

CURRENT THROUGH LOAD

(c)

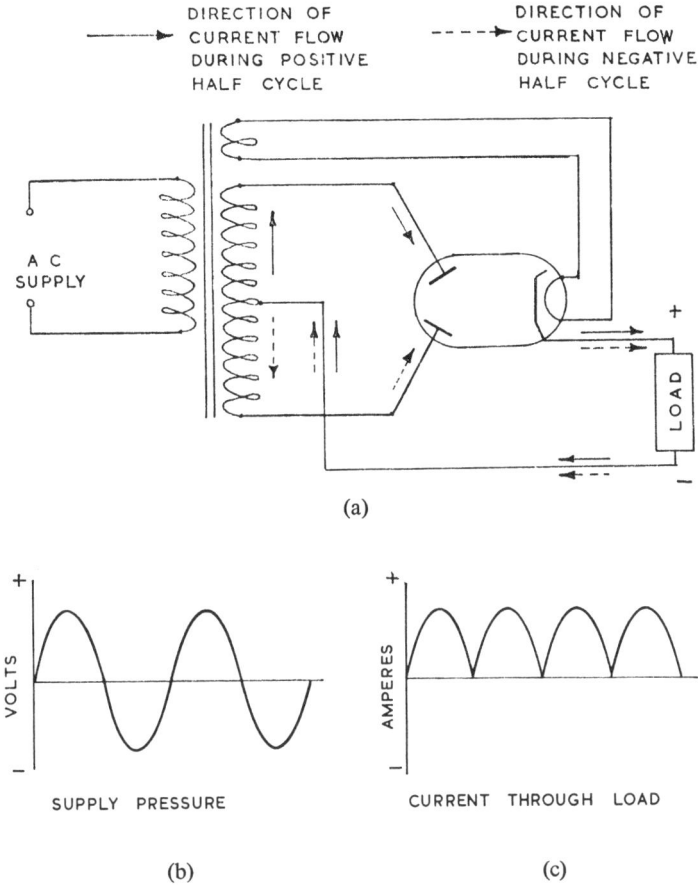

FIGURE 10.26   Full-Wave Rectification Using Thermionic Diode Valve.
(a) Circuit Diagram.
(b) Waveform of Supply e.m.f.
(c) Waveform of Load Current.

10.72 Figure 10.27 shows a half wave rectifier circuit supplying a resistive load. Smoothing is provided by the capacitor, C, connected in parallel with the load. Current flows through the diode only when the input voltage is more positive than the upper terminal of the capacitor. This current will be used partly to supply the load but mainly to charge the capacitor up to the maximum value of the supply voltage. During the period when no current flows through the diode, the capacitor discharges through the load so maintaining the load current. The p.d. across the capacitor (and load) falls slightly during this period until, after one cycle of the a.c. waveform, current again flows through the diode and the

239

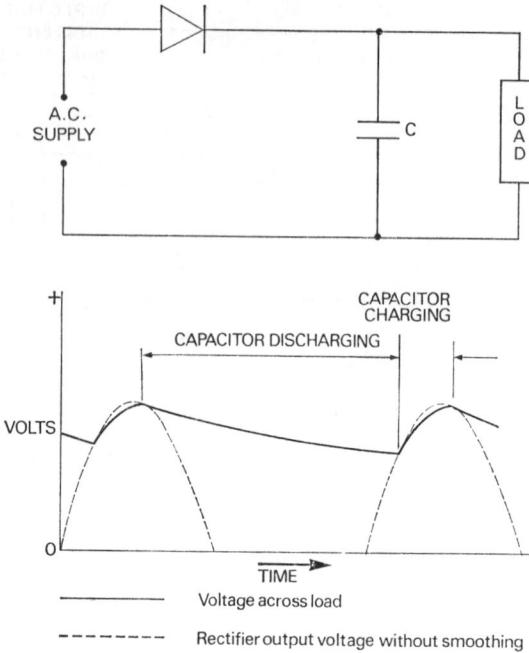

FIGURE 10.27   Simple Capacitor Smoothing.
(a) Circuit diagram.
(b) Voltage waveforms.

capacitor is recharged. The resulting voltage waveform across the load is shown in Fig. 10.27(b) it can be seen that the voltage fluctuation is considerably reduced. One disadvantage of this arrangement is that the capacitor must gain sufficient charge during the time that the diode conducts to replace the charge used in maintaining the load current for the remainder of the time. The more constant the capacitor maintains the voltage, the shorter is the conduction period of the diode, as it can conduct only while the supply voltage is higher than the capacitor voltage. Thus increasing the capacitance to obtain better smoothing also increases the diode current, which can ultimately result in damage to the diode.

10.73 Another method of obtaining smoothing is to connect an inductor in series with the load as shown in Fig. 10.28. Whenever the current in an inductor changes, an e.m.f. is induced in it which opposes the change, see Section 7.6. This induced e.m.f. will oppose the voltage output from the diode when the current is rising, but will add to the rectifier output when the current is falling, thus producing a much more constant voltage across the load.

10.74 Better smoothing can be obtained by making use of both inductance and capacitance. Figure 10.29 shows a typical circuit. Although

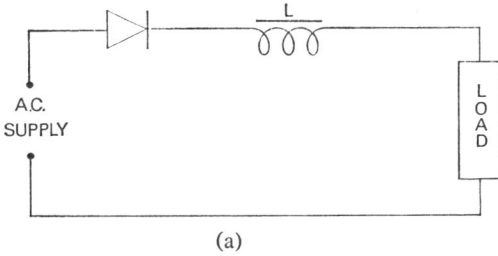

(a)

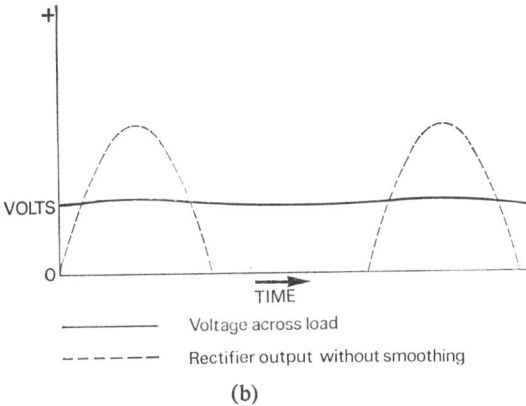

TIME

——————— Voltage across load

– – – – – – Rectifier output without smoothing

(b)

FIGURE 10.28   Simple Inductance Smoothing.
(a) Circuit Diagram.
(b) Voltage waveforms.

the action of the smoothing circuits above has been shown as being supplied by a half-wave rectifier, they are all equally suitable for use with full-wave rectifiers. Indeed, the full wave circuit will give a better performance as it provides two driving impulses per cycle in place of the one impulse per cycle produced by the half-wave circuit.

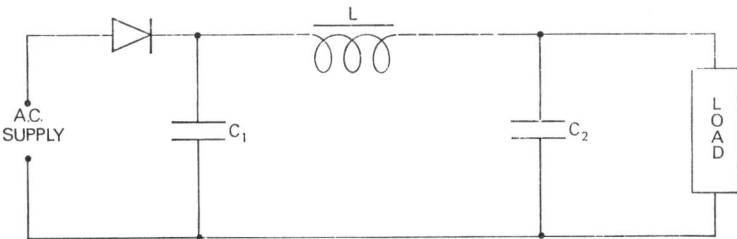

FIGURE 10.29   Capacitance Input Smoothing Circuit.

241

SUMMARY OF FORMULAE

Transformers:
$$\begin{cases} \dfrac{E_p}{E_s} = \dfrac{N_p}{N_s} \\[2mm] \dfrac{I_p}{I_s} = \dfrac{N_s}{N_p} \end{cases}$$

$$\text{Regulation} = \frac{\text{Volts on no--load} - \text{Volts on load}}{\text{Volts on no--load}}$$

Speed of Rotating Magnetic Field: $\quad N_s = \dfrac{60 \times 2f}{p}$

Induction Motor: Fractional Slip: $\quad = \dfrac{N_s - N_r}{N_s}$

EXERCISE 10

1. Describe, with the aid of sketches, the operation of a single-phase, double-wound transformer.
   The primary winding of a step-down, single-phase transformer takes a current of 9·09 A at 11 000 V when working at full load. If the transformation ratio is 1·67:1, calculate the secondary voltage and current. W.J.E.C.

2. Describe the operation of a single-phase transformer, explaining clearly the effects of the various parts.
   The primary winding of a step-down single-phase transformer takes a current of 22 amperes at 3300 volts when working at full load. If the transformation ratio is 14 to 1, calculate the secondary voltage and current. C.G.L.I.

3. Make a neat diagram or sketch of a simple single-phase double-wound transformer, and with its aid explain the action of the transformer.
   Calculate the respective numbers of turns in each winding of such a transformer which has a step-down ratio of 3040 V to 240 V if the "volts per turn" are 1·6.

4. Describe briefly, with the aid of sketches, how an auto-transformer differs from a double-wound transformer. Give one advantage and one disadvantage of the auto-transformer compared with a double-wound transformer.
   A single-phase auto transformer is supplied at 220 V, 50 Hz, the supply being connected between the common terminal and a tapping at 720 turns. Calculate the number of turns required to give an output of 240 V and draw a diagram of connections for the transformer indicating the number of turns in each part of the winding.

5. Explain carefully how a three-phase a.c. supply can be used to produce a rotating magnetic field.
   A 415 V, 50 Hz, three-phase a.c. supply is used to produce a 6-pole rotating field. At what speed does the field rotate?

6. Explain the terms *slip* and *synchronous speed* as applied to an induction motor.
   A 4-pole, 3-phase induction motor is connected to a 415 V, 50 Hz, 3-phase supply. When fully loaded the motor runs at 1420 rev/min. What is the full load percentage slip?

7. What do you understand by the *slip* of an induction motor?
    Calculate the speed in rev/min of a 6-pole induction motor which has a slip of 6 per cent at full load, with a supply frequency of 50 Hz. What will be the speed of a 4-pole alternator supplying the motor? C.G.L.I.
8. Explain the meaning of "rectification" of an alternating current.
    Describe briefly, with sketches, one form of single rectifier, and give a diagram of connections. C.G.L.I.
9. A direct current supply is needed to charge a number of secondary cells. The mains supply is alternating current, single phase. Describe briefly *two* methods of providing the direct current. C.G.L.I.
10. Show by means of diagrams what you understand by *rectification* of an alternating current.
    Describe with sketches one form of metal rectifier element and show how you would connect four such elements to give "full wave" rectification. C.G.L.I.
11. With the aid of suitable diagrams explain how it is that current can flow in only one direction through a thermionic diode valve. How can such a valve be used to provide a direct current output from an alternating current supply?
12. What do you understand by the following terms as applied to a thermionic diode valve:
    (a) Anode
    (b) Cathode
    (c) Electron.
    Draw a circuit diagram showing how two single diode or one double-diode valve(s) should be connected to provide full-wave rectification. Sketch wave forms of the a.c. input and the output voltages of this circuit.

11 Electrical Measurements

11.1 Electrical Instruments

11.11 Instruments for measuring electric currents may take advantage of any of the effects of an electric current. For example, a measure of the size of a current could be obtained by the heat it produces when flowing through a wire, or the amount of chemical change it produces when flowing through an electrolyte. In practice the majority of electrical instruments in everyday use make use of the magnetic effect of a current. The moving coil meter utilises the torque produced when a current flows through a coil situated in a magnetic field (Section 7.2). The moving iron meter uses the repulsion effect between like magnetic poles, the strength of the poles being governed by the value of the current flowing in a coil. Electrical pressures are usually determined by measuring the current which they produce when applied to a known, high-value resistor, although use can be made of the electrostatic attraction which exists between oppositely charged surfaces.

11.12 There are three important factors which must be taken into account when studying the operation of an electrical instrument. Firstly, there is the method of producing the *operating torque*, which drives the meter pointer around the scale. Secondly, there is the *control* or *restoring torque*; this torque balances the operating torque when the instrument pointer is at the correct deflection; it could be said that, in effect, the control torque weighs the operating torque. In most modern instruments the control torque is provided by using hair springs. Finally *damping* is provided to prevent the pointer from swinging wildly about whenever it is called upon to respond to a sudden change in the current being measured. When the pointer is required to move suddenly to a new reading it has a natural tendency to overshoot the correct mark, and then to oscillate backwards and forwards about the correct mark for some time before finally coming to rest. Ideally the damping should prevent the pointer from over-swinging, while not restricting the freedom of the pointer to move. When this result is achieved the meter pointer moves quickly to its correct indication and then stays there until the current again changes. An instrument which does this is said to be *dead beat*.

11.2   The Moving Coil Instrument

11.21   Figure 11.1 shows the construction of a moving coil instrument. The moving coil itself is rectangular in shape, and consists of a number of turns of fine insulated copper wire wound on a light aluminium frame. The coil is supported by spindles resting in jewelled bearings so that it is free to rotate in the gaps between the soft iron pole pieces and the fixed soft

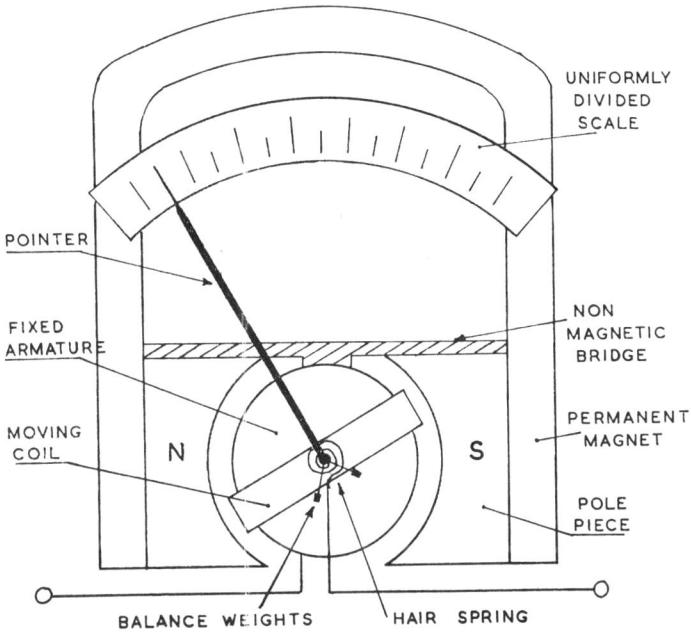

FIGURE 11.1   Moving Coil Instrument.

iron armature. The magnetic field in the gaps is produced by a powerful permanent magnet. The fixed soft iron armature, which is supported centrally between the pole pieces, serves to concentrate the magnetic flux so that a uniform magnetic field is obtained in the gaps. The current to be measured is led into, and out of, the coil via two phosphor bronze hair springs, which are fitted one at each end of the coil. When the coil rotates the hair springs will either wind up or unwind, in either case they will produce a torque tending to return the coil to the zero position, and so the hair springs also provide the control torque for the meter. A light pointer attached to the coil indicates the value of the current on a uniformly divided scale. The weight of the pointer is counter-balanced by two or more small balance weights, so that there is no tendency for the weight of

245

the pointer to cause the coil to rotate and so give a false reading. Whenever the coil rotates, eddy currents are induced in the aluminium frame on which the coil is wound. These currents produce a braking effect so providing the necessary damping torque.

11.22 The direction of the torque produced by the moving coil depends on the direction of the current through the coil. Thus the instrument is not suitable for measuring alternating currents as, if an alternating current were passed through the coil, it would merely attempt to vibrate backwards and forwards. The coils of most instruments cannot move backwards and forwards quickly enough to respond to a current at mains frequency, so no indication at all is obtained when an alternating current from the mains is passed through them, even though the current may be high enough to burn out the coil. Owing to the delicate construction of a moving coil instrument it must always be handled with care. The basic instrument is only suitable for measuring very small currents, up to about 50 milliamperes. Methods of adapting the basic instrument for measuring higher currents and for measuring voltages are discussed in Section 11.3.

11.3 Range Extension of Moving Coil Instruments

11.31 It is necessary to pass only a very small current through the coil of a moving coil instrument in order to drive the pointer to full scale deflection. In order to measure a larger current it is necessary to by-pass most of the current, allowing only a small known fraction to pass through the coil of the instrument. This is achieved by connecting a low value resistor in parallel with the moving coil, such a resistor being called a *shunt* resistor. A voltage can be measured by determining the value of the current that it causes when applied to a circuit of known resistance. Thus a voltmeter can be constructed by connecting a high value resistor in series with the moving coil; such a resistor is sometimes called a *multiplier*.

11.32 The value of the shunt resistor required to adapt a given moving coil movement so that it gives full scale deflection for any required current can be easily calculated using Ohm's law, as is shown by the following example.

Example 11.1

A moving coil instrument has a coil of resistance 8 Ω and a current of 10 mA through the coil gives full scale deflection. Determine the value of the shunt required so that the instrument will have a range of 0–2 A.

Solution

In order to use the given instrument as an ammeter, a shunt resistor must be connected in parallel with it as shown in Fig. 11.2(a). The electrical circuit then consists of two resistors in parallel as shown in Fig. 11.2(b), the 8 Ω resistor being the instrument coil and R_s being the shunt. The

246

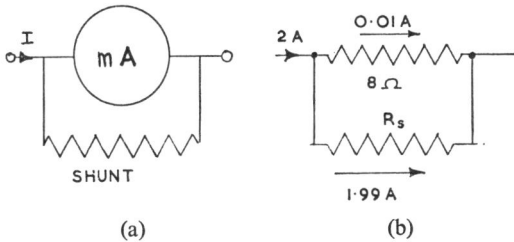

FIGURE 11.2 Example 11.1.
(a) Shunt Connected in Parallel with Milliammeter.
(b) Equivalent Electrical Circuit.

meter is required to give full scale deflection for a total current of 2 A. In order to produce full scale deflection the current in the coil has to be 10 mA (or 0·01 A) so that the amount of current to be by-passed by the shunt is:

$$I = 2 - 0·01$$
$$I = 1·99 \text{ A}$$

These values have been marked on Fig. 11.2(b).

Now by Ohm's law:

p.d. across meter coil, $V = RI$
$$V = 8 \times 0·01$$
$$V = 0·08 \text{ V}$$

As R_s is connected in parallel with the meter coil this must also be the p.d. across R_s, so, using Ohm's law again:

$$R_s = \frac{V}{I}$$
$$R_s = \frac{0·08}{1·99}$$
$$R_s = 0·0402 \text{ ohms.}$$

11.33 The value of the series resistor required to adapt a given moving coil movement so that it gives full scale deflection for any required voltage can also be calculated using Ohm's law, as in the following example.

Example 11.2

A moving coil instrument having a coil of resistance 100 Ω requires a current of 1 mA to give full scale deflection. Determine the value of the series resistor required so that the instrument will have a range of 0–200 V.

Solution

In order to use the given instrument as a voltmeter, a resistor must be connected in series with it as shown in Fig. 11.3(a). The electrical circuit

247

then consists of two resistors in series as shown in Fig. 11.3(b), the 100 Ω resistor being the instrument coil and R_m being the multiplier.

When a p.d. of 200 V is applied to the instrument terminals, a current of 1 mA must flow through the moving coil in order to give full scale

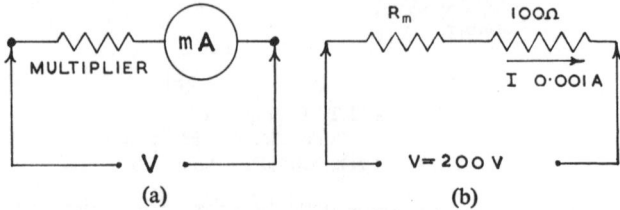

FIGURE 11.3   Example 11.2.
(a) Multiplier Connected in Series with Milliameter.
(b) Equivalent Electrical Circuit.

deflection. It follows that the total resistance of the instrument circuit must be:

$$R_T = \frac{V}{I}$$

$$R_T = \frac{200}{0\cdot001}$$

$$R_T = 200\ 000\ \Omega$$

As the coil itself contributes a resistance of 100 Ω to this total it follows that:

Value of series resistor, $R_m = R_T - R$ coil

$$R_m = 200\ 000 - 100$$

$$\underline{R_m = 199\ 900\ \Omega}$$

11.4   Moving Iron Meters

11.41   Figure 11.4 shows the principle of the moving iron meter. When an electric current flows through the coil it magnetises the fixed and moving irons with like polarities. As like poles repel each other, the moving iron is repelled from the fixed iron, so turning the spindle and causing the pointer to move over the scale. The scale is not uniformly divided, the divisions tending to be cramped at the lower end of the scale. The control torque is usually provided by a phosphor bronze hair spring but, in some cases, a small weight is fastened to a lever arm attached to the spindle. Rotation of the spindle raises this weight, which will then exert a moment

248

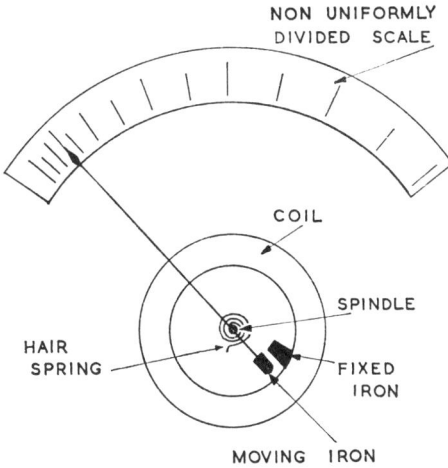

FIGURE 11.4   Moving Iron Meter.

tending to return the spindle to the zero position. This method of providing the control torque is known as gravity control. Damping is provided by an air dashpot which is not shown in Fig. 11.4 but is illustrated separately in Fig. 11.5. A light aluminium vane, attached to the meter spindle moves inside a close fitting sector shaped box. The sides of the vane are very close to, but do not actually touch the sides of the box. When the meter spindle rotates, the movement of the vane displaces the air from one side of the box to the other. The movement of this air is restricted because of the very small clearance between the sides of the vane and the box, and this provides the required damping force.

11.42   If the direction of the current in the coil is reversed, the polarities of both fixed and moving irons are also reversed. However both the irons

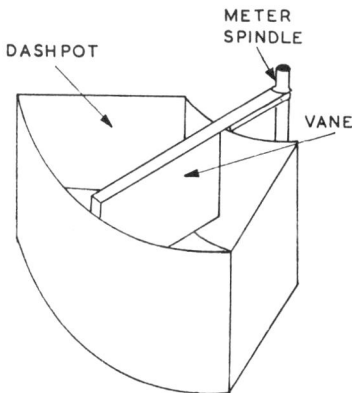

FIGURE 11.5   Air Dashpot Damping.

still have the same polarity as each other, and so the meter produces a deflection in the same direction as before. Thus when an alternating current is passed through the coil the instrument deflection is in the same direction for each half cycle of the waveform and a steady indication is obtained. It can be shown that the reading obtained gives the r.m.s. value of the alternating current, which is the value normally required. The coil of a moving iron instrument can be wound with heavy gauge wire as it is fixed in position, the number of turns in the coil being chosen to suit the current range required. Moving iron voltmeters use a coil of many turns of fine wire, often with a resistor in series.

11.5 Further Applications of Moving Coil Instruments

11.51 Although in Section 11.22 it was stated that the moving coil instrument was not suitable for measuring alternating currents, the

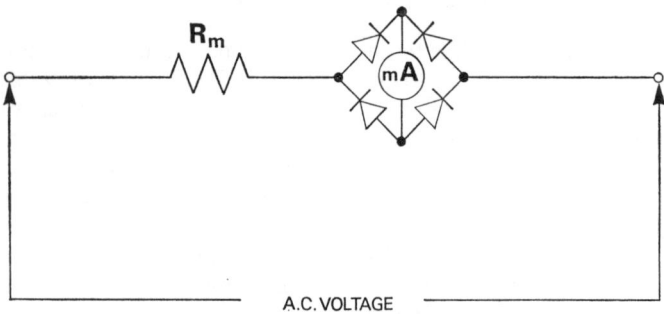

FIGURE 11.6   Rectifier Moving Coil Voltmeter.

versatility of this instrument can be increased by the addition of a rectifier to convert an a.c. input into a d.c. current suitable for operating the meter. Figure 11.6 shows the circuit of a rectifier type moving coil voltmeter suitable for measuring a.c. voltages. The resister R_m serves the same purpose as the multiplier used in the d.c. moving coil voltmeter previously described, but the current flowing through this resistor is converted into d.c. by the bridge rectifier circuit shown, before flowing through the meter coil. If the value of the multiplier, R_m, is calculated in the same way as for d.c. voltmeters, then the instrument will indicate the mean value of the a.c. voltage applied to it. It is common practice to reduce the value of R_m by 11 % from its d.c. value, the result of which is to multiply the meter reading by 1.11 which is the form factor of a sine wave. By doing this the meter will correctly indicate the r.m.s. value of an a.c. voltage, provided that it has a sinusoidal waveform; the instrument is said to be calibrated to read *r.m.s. values of a sinewave*. It must be remembered that such
250

instruments can have large errors when measuring waveforms which are not sinusoidal.

11.52 Figure 11.7 shows how a moving coil meter can be adapted to measure alternating currents. Again a bridge rectifier is needed to rectify the current, for a half wave type of rectifier would be unsuitable as it would prevent the flow of current during alternate half cycles and an ammeter should never impede the flow of the current which it is measuring. A transformer is used in place of the shunt resistor used in d.c. ammeters. This transformer is called a *current transformer* and has several advantages. The current ratio is the reciprocal of the turns ratio and is accurately determined by winding the transformer with the correct number of turns. This normally means that the input or primary winding needs far fewer turns than the secondary; for heavy currents a single turn winding often suffices. It follows that the p.d. needed across the primary winding is usually far less than the p.d. across the moving coil unit, which is in any

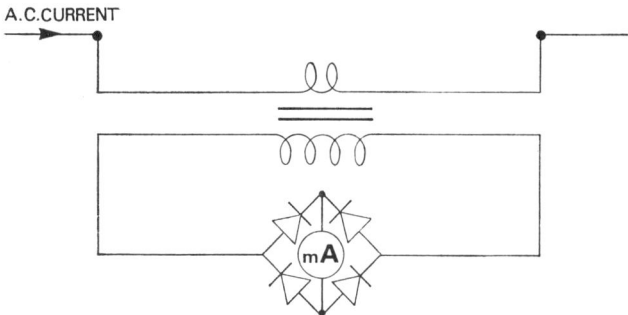

FIGURE 11.7 Rectifier Moving Coil Ammeter.

case quite small. Thus the meter can have a very low input impedance which means that it causes only a very slight disturbance to the circuit in which it is connected.

11.53 The rectifier moving coil principle is obviously extremely convenient for use in multi-purpose multi-range test instruments. In such instruments a single moving coil instrument movement provides all the indications. Switches are used to select the correct shunts, multipliers, rectifiers etc. required for the range in use. The rectifier meter has an advantage for some situations in that it usually requires less current to operate than a moving iron meter, i.e. it is more sensitive. Also, whereas most moving iron meters can read accurately only up to a frequency of about 100 Hz, rectifier type meters can be made accurate for frequencies up to 20 kHz or more.

11.54 If a junction is made between two dissimilar metals, an e.m.f. is produced whenever the junction is heated. This is called the *thermo-electric effect*, the junction of the two metals being called a *thermo-couple*.

Thermocouples can be employed as thermometers, and they can also be used as ammeters. In order to use the thermocouple as an ammeter the current to be measured is caused to heat the thermocouple element and the e.m.f. which is then developed by the thermocouple is used to operate a moving coil instrument. As the meter depends for its operation on the heating effect of the current being measured, it automatically indicates r.m.s. values. The thermocouple instrument is suitable for use up to very high radio frequencies. Figure 11.8 illustrates the principle of the thermocouple instrument.

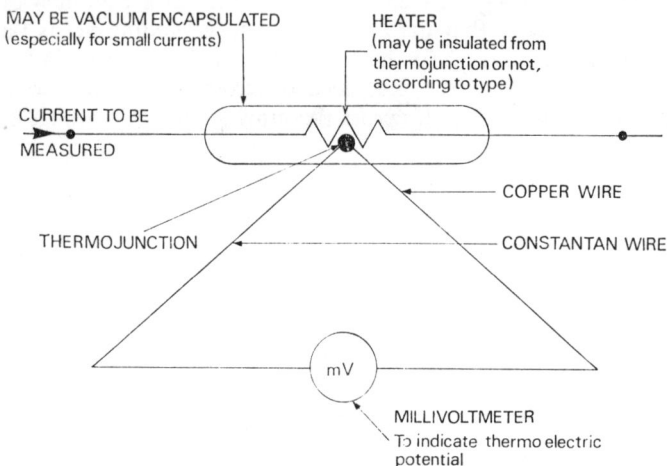

FIGURE 11.8 Principle of Thermocouple Ammeter.

11.55 Transistor or valve amplifiers can be added to moving coil instruments to increase their usefulness. It is beyond the scope of this book to discuss the details of the many electronic circuits which have been developed to improve the sensitivity and frequency response of the basic moving coil indicator.

11.6 Other Instrument Types

11.61 Figure 11.9 illustrates the electrostatic voltmeter. The operating torque of this instrument is caused by the electrostatic forces set up between the fixed and moving vanes whenever there is a potential difference between them. The electrostatic meter usually has hair springs to provide the control torque and an air dashpot or aluminium disc and magnet to provide damping. This type of meter is more suitable for measuring high voltages, as low voltages produce only a very small electrostatic force, so that a low voltage instrument tends to be delicate and expensive. The
252

FIXED VANES A
Connected to one terminal

FIXED VANES C AND
MOVING VANES B
Connected to the other
terminal

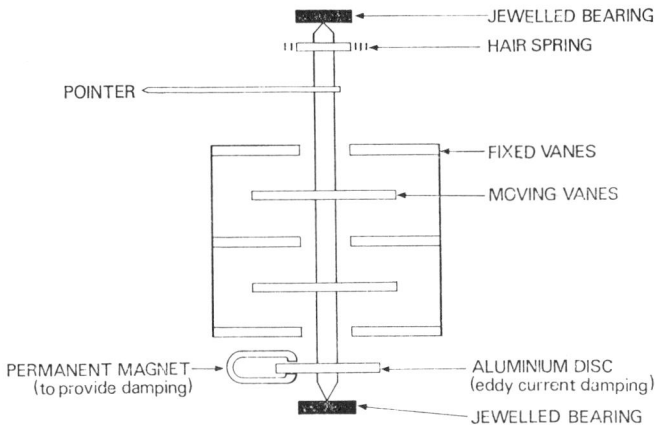

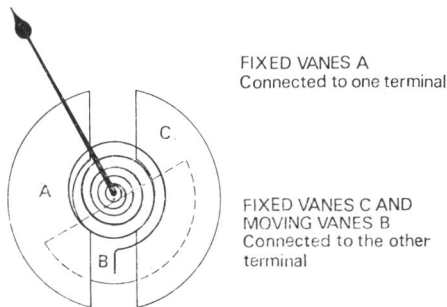

FIGURE 11.9   Electrostatic Voltmeter.

instrument has the advantage when used to measure d.c. voltages that it consumes no current at all. The electrostatic voltmeter can also be used to measure a.c. voltages up to high frequencies, in this case the instrument is equivalent to a small capacitor and so draws some current from the circuit being tested.

11.62   Figure 11.10 illustrates the principle of the dynamometer or electrodynamic type wattmeter. In this instrument the current flowing in the circuit is passed through a pair of fixed *current coils*, which produce a magnetic field proportional to the current. The voltage acting on the circuit is applied to the moving coil, or *potential coil*, via a multiplier resistor R_m so that the current through this coil is proportional to the supply voltage. Thus the torque developed depends on both the current and the voltage, that is it is proportional to the power consumed by the circuit. The control torque of the wattmeter is normally provided by hair springs, and the damping by means of an air dashpot. By connecting

253

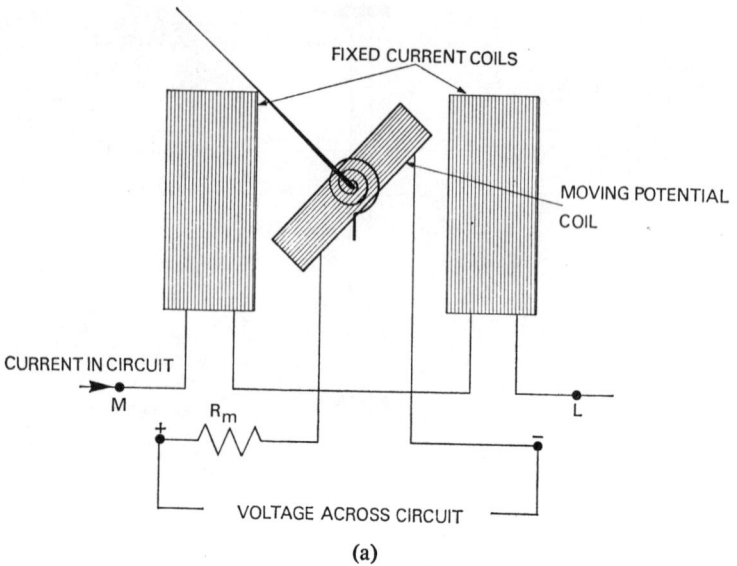

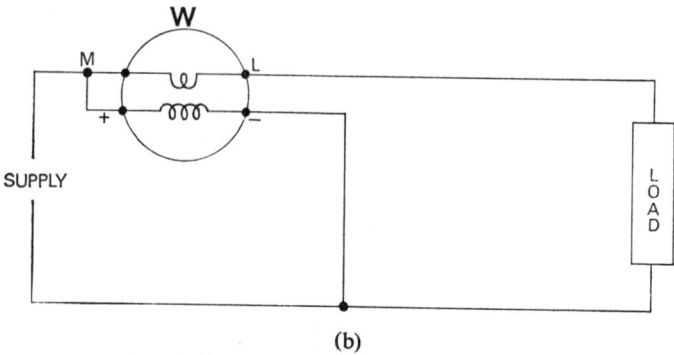

FIGURE 11.10 Principle of Dynamometer Wattmeter.
(a) Arrangement of coils.
(b) Wattmeter connected to measure power consumed by a load.

the fixed and moving coils in series it is possible to adopt the dynamometer instrument for measuring either voltage or current. For these purposes the current coils are normally designed to operate with a very small current. A series resistor is used to provide voltage ranges, as in the moving coil meter, and current ranges can be provided by using shunts for a.c./d.c. instruments or current transformers in a.c. only instruments.

11.63 Figure 11.11 shows the principle of the ohmeter. The instrument has two moving coils, A and B, current being led into, and out of, the coils

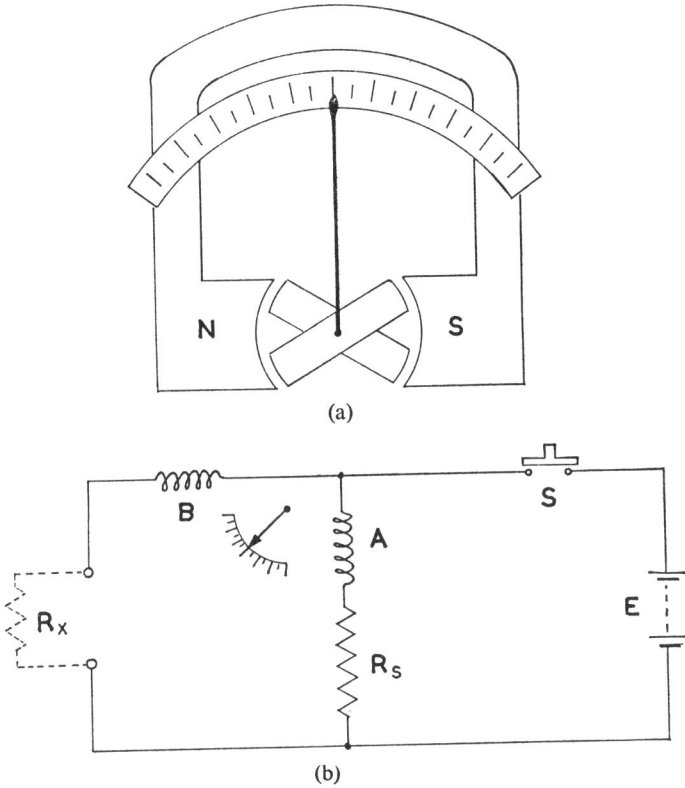

FIGURE 11.11   Ohmeter.
(a) Arrangement of coils.
(b) Electrical circuit.

by means of fine ligaments which do not restrain the movement of the coils in any way. Figure 11.11(a) shows how the coils are pivoted so as to rotate freely within the magnetic field provided by a permanent magnet. Figure 11.11(b) shows the electrical connections of the instrument. The coils are so connected that the currents through them tend to make them rotate in opposite directions. The instrument is provided with a press button switch S. In order to test an unknown resistor R_x, the button is pressed. The current which then flows through coil A, known as the *control coil*, is governed by the value of the internal resistor R_s, while the current which flows through coil B, known as the *deflecting coil* is governed by the value of the unknown resistor R_x. If R_x has a low value, the current in coil B is greater than that in coil A, and the pointer moves towards the zero end of the scale. On the other hand if R_x has a high value, coil A has the higher current and the pointer moves towards the infinity end of the scale. If R_x

255

and R_s have equal values, the coils have equal currents and the pointer takes up a position at the middle of the scale. Thus the position of the pointer depends on the value of R_x and the scale is marked in ohms.

11.64 The battery driven instrument is suitable for measuring low values of resistance and is often used as a *continuity tester*. If the instrument is connected between the two ends of a circuit, a low reading indicates that the circuit is continuous, as a continuous conductor has a low resistance. If however there is a break in the circuit, the resistance becomes infinite, and the instrument will indicate accordingly. When high resistance values, of the order of megohms are to be measured, for example when testing the insulation resistance of an installation, the e.m.f. produced by a small battery is inadequate. A common form of insulation tester uses a hand-cranked generator to provide the necessary high testing voltage. The circuit of such an instrument is shown in Fig. 11.12. It will be noticed that

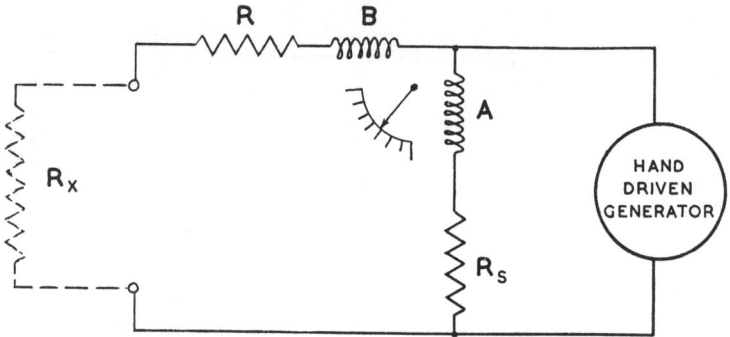

FIGURE 11.12   Circuit of Insulation Resistance Tester.

an additional resistor R is shown in series with the deflecting coil; this helps to protect the instrument against excessive current if it is inadvertently connected to a live circuit. Some instruments also incorporate a fuse in this position to provide extra protection. Other types of insulation tester are

Uses of Meters	
Type of Meter	*Used to Measure*
Moving coil	d.c. voltage and current
Moving coil + rectifier	a.c. voltage and current up to high frequencies
Moving coil + thermocouple	a.c. current up to very high frequencies
Moving iron	d.c. voltage and current
	a.c. voltage and current at power frequencies
Electrostatic	d.c. and a.c. voltage up to high frequencies
Dynamometer	d.c. and a.c. voltage, current and power.
Ohmeter	resistance

available, in which the required high value of test voltage is produced from a low voltage battery either by means of an induction coil, or by a transistorised voltage converter.

11.65 The chart at the bottom of page 256 compares the uses of the various types of meter discussed in the previous sections.

11.7 Errors due to Meter Connections

11.71 Indicating instruments normally consume some power in order to provide the torque required to move the pointer. Although this power is usually small, it has to be provided by the circuit being tested and may itself affect the operation of the circuit, so giving rise to incorrect readings.

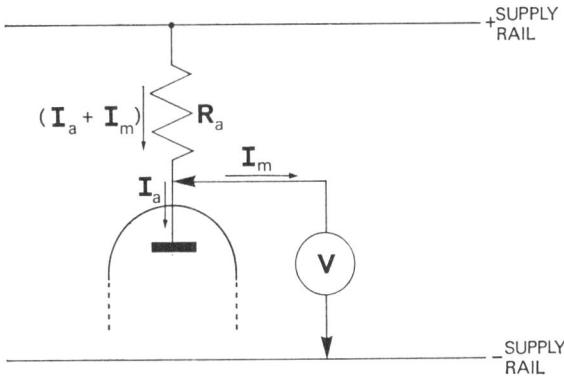

FIGURE 11.13 Measuring the Anode Voltage of Valve.

These effects seldom give rise to serious difficulties in power circuits, but can have far reaching results when testing the light current circuits which are encountered in the field of electronics testing. For example, Fig. 11.13 shows part of the circuit of a triode amplifying circuit in which a voltmeter, V, is being used to measure the anode voltage of the valve. The voltmeter draws a current I_m and this current must flow through the load resistor, R_a, in addition to the normal anode current, I_a. This results in an additional voltage drop in R_a so that the anode voltage measured is lower than its normal value. If the meter current, I_m, is of the same order as the anode current, I_a, then very serious errors can result. In high frequency circuits even the stray capacitance and inductance of long meter leads may make readings meaningless unless due precautions are taken. It is for this reason that many high frequency instruments are provided with special leads and probes for connecting them to the circuit to be tested. Electronic instruments, such as valve voltmeters are very useful in these circumstances as such instruments usually draw only a very

257

small current from the source being tested, i.e. they have a very high sensitivity.

11.72 In some cases the error caused by instrument connections can easily be allowed for, as illustrated by the following example.

Example 11.3

An ammeter having a resistance of $0.5\ \Omega$ and a voltmeter having a resistance of $2\ 000\ \Omega$ are used to measure the value of two resistors. In order to measure the first resistor, the circuit of Figure 11.14(a) was used, the readings obtained being $V = 10$ V, $I = 0.05$ A. In order to measure the second resistor the circuit of Figure 11.14(b) was used, the readings

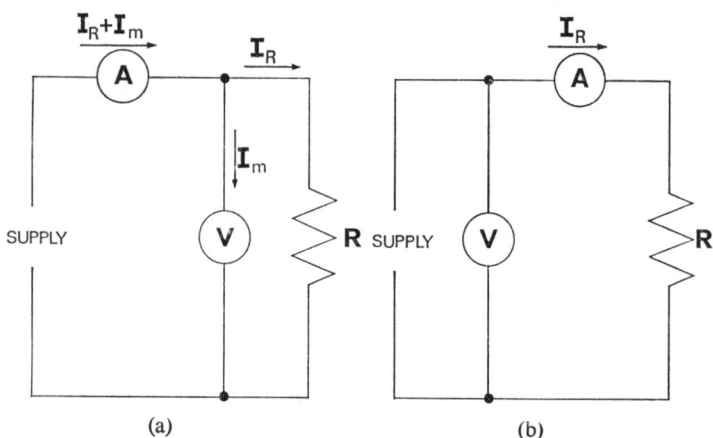

FIGURE 11.14 Example 11.3.

obtained being $V = 5$ V, $I = 1.0$ A. In each case calculate the true and apparent resistance and hence deduce the percentage error caused by the meter connections. Criticise each circuit and suggest an improved method of connection.

Case 1

$$\text{Apparent Resistance} = \frac{\text{Voltmeter reading}}{\text{Ammeter reading}}$$

$$R = \frac{10}{0.05}$$

$$\underline{R = 200\ \Omega}$$

This result is in error because the voltmeter current, I_m, flows through the ammeter in addition to the current I_R.

$$I_m = \frac{V}{R_m}$$

$$I_m = \frac{10}{2000}$$

$$I_m = 0.005\,\text{A}$$

$$\therefore\quad I_R = 0.05 - 0.005$$

$$I_R = 0.045\,\text{A}$$

The true value of R is therefore given by:

$$\text{True Resistance} = \frac{V}{I_R}$$

$$R = \frac{10}{0.045}$$

$$\underline{R = 222.2\,\Omega}$$

The error caused by the instrument connections is seen to be 22.2 Ω; expressing this as a percentage of the true value:

$$\%\,\text{error} = \frac{22.2 \times 100}{222}$$

$$\%\,\text{error} = 10\%$$

Criticism

The circuit used is not suitable for measuring high value resistors. For this purpose the circuit of Fig. 11.14(b) would be far more suitable as the error is only 0.5 Ω due to ammeter resistance as explained in Case 2 below.

Case 2

$$\text{Apparent Resistance} = \frac{\text{Voltmeter reading}}{\text{Ammeter reading}}$$

$$R = \frac{5}{1}$$

$$\underline{R = 5\,\Omega}$$

This result is in error because in this case the voltmeter registers the p.d.

259

across the ammeter's resistance and R combined. Thus the true value of R is given by:

True resistance = Apparent resistance − Ammeter resistance

$$R = 5 - 0.5$$
$$\underline{R = 4.5\ \Omega}$$

The error is seen to be $0.5\ \Omega$; expressing this as a percentage of the true value:

$$\%\ \text{error} = \frac{0.5 \times 100}{4.5}$$
$$\underline{\%\ \text{error} = 11.1\%}$$

Criticism

The circuit used is not suitable for measuring low value resistors. For this purpose the circuit of Fig. 11.14(a) would be more suitable as the error would then be due to a voltmeter current of only 0.0025 A, which is small compared with the resistor current of 1.0 A.

11.8   Combinations of Instruments

11.81   It is sometimes necessary to use a combination of instruments in order to measure a particular quantity. For example it may be necessary to determine the power factor of some apparatus connected to an a.c. supply. This could be done using a power factor meter if this type of meter were available, but if no power-factor meter were to hand then a good practical alternative method would be to use a combination of instruments

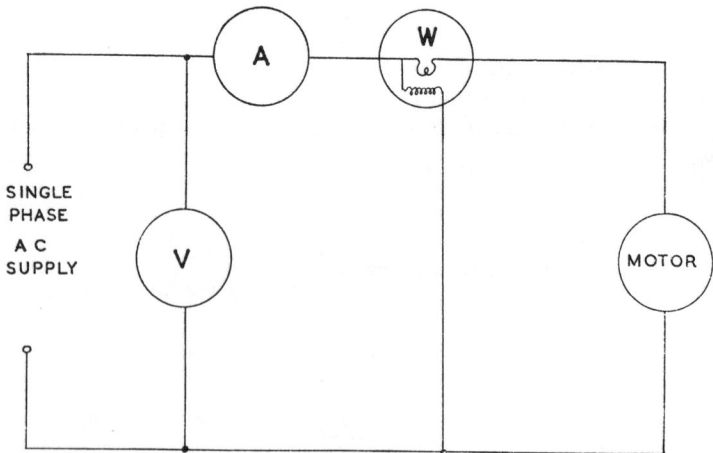

FIGURE 11.15   Connections for Measuring Power Input to Single Phase a.c. Motor.

to measure the power, current and supply pressure, the power factor being then determined using the formula:

$$\text{power factor} = \frac{\text{True power}}{\text{Apparent power}}$$

Figure 11.15 shows the connections for determining the power consumed and power factor of a single phase a.c. motor. Using this arrangement of instruments the following information may be obtained:

Power input = Wattmeter reading (in kW)

$$\text{Input in kV A} = \frac{\text{Ammeter reading} \times \text{Voltmeter reading}}{1000}$$

$$\text{Power factor} = \frac{W}{VA} \quad \text{or} \quad \frac{kW}{kVA}$$

Example 11.4

Make a diagram showing the connections of voltmeter, ammeter and wattmeter, in a single-phase a.c. circuit supplying power to a motor.

The following values were recorded in a load test of a single-phase motor. Calculate the efficiency of the motor and its power factor:

$$\begin{aligned}
&\text{Voltmeter reading} \ldots \ldots 240 \text{ V} \\
&\text{Ammeter reading} \ldots \ldots \ 75 \text{ A} \\
&\text{Wattmeter reading} \ldots \ldots 13 \text{ kW} \\
&\text{Mechanical output} \ldots \ldots 11 \cdot 2 \text{kW}
\end{aligned}$$

Solution

The connection diagram required is shown in Fig. 11.15.

Power input to motor = 13 kW

Power output from motor = 11·2 kW

$$\text{Efficiency} = \frac{\text{Output}}{\text{Input}} \times 100\%$$

$$\text{Efficiency} = \frac{11 \cdot 2 \times 100}{13}$$

$$\underline{\text{Efficiency} = 86 \cdot 1\%}$$

$$\text{Power factor} = \frac{W}{VA}$$

$$\text{p.f.} = \frac{13\,000}{240 \times 75}$$

$$\underline{\text{p.f.} = 0 \cdot 722}$$

261

11.9 Electrical Energy Measurements

11.91 Electrical energy is measured either in joules or kilowatt hours (Chapter 1, Section 1·6). The electrical energy consumed by an installation is measured by a kilowatt hour meter. This instrument is essentially a wattmeter in which the moving element, often an aluminium disc, is free to rotate. The speed at which the moving element rotates is proportional to the power being used. As the number of revolutions made by the moving element depends both on its speed and on the time for which it revolves, the total number of revolutions is proportional to the amount of electrical energy which has passed through the meter. The revolutions are indicated by a *counting train* geared to the rotating element, the dials of this counting train being marked to indicate the energy consumed in kW h. Figure 11.16 shows the method of connecting a kilowatt hour meter.

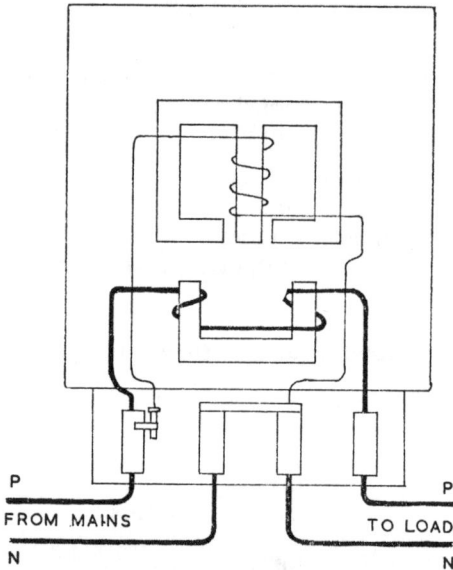

FIGURE 11.16   Connections of kW h Meter.

11.92 Figure 11.17 shows the dials of a typical kilowatt hour meter. The reading is indicated by pointers which are geared to move continuously, thus any pointer may well be in between two scale points, in which case the correct indication is given by the figure which the pointer has just passed. In some cases it is difficult to decide whether the pointer has actually passed a given figure or not. It is then necessary to observe the next lower dial; if the pointer on this dial has not yet reached its zenith then the pointer in doubt has not yet passed its scale mark. In Fig. 11.17, the 100's pointer reading is doubtful until it is noted that the 10's pointer indicates approximately 1 showing that the 100's pointer has in fact passed the 2 mark. The 10's indication is itself in doubt until it is observed that the 1's pointer indicates 8, showing that the 10's pointer is not yet at 1 and the correct reading for this dial is 0. The complete indication of the meter represented

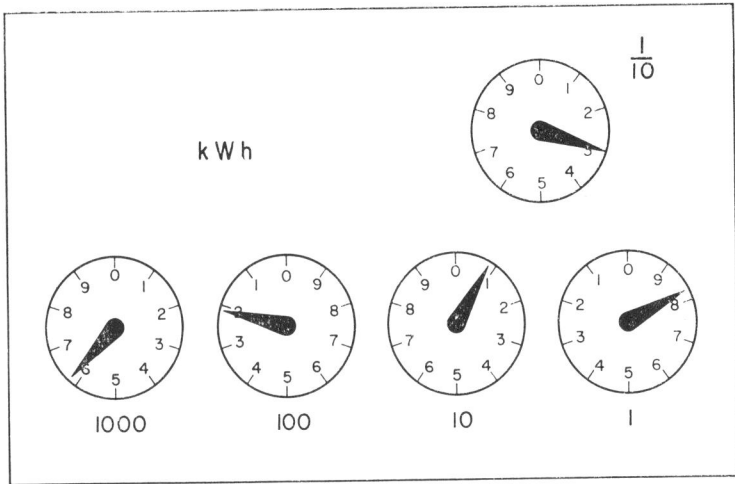

FIGURE 11.17   Dials of kW h Meter.

in Fig. 11.17 is 6208·3 units. In practice, to obtain a consumer's consumption over a period, the meter reading at the beginning of the period is subtracted from the reading at the end of the period.

11.10 Tariffs

11.101   The costs of supplying electricity to a consumer fall into two main categories;
   (a) Fixed costs, which must be met irrespective of how much electricity is actually supplied. These costs include such items as, interest on capital, depreciation of plant, salaries, taxes and rents.
   (b) Running costs, which depend on the number of units of electricity actually supplied. These costs include such items as fuel, and some of the salaries and maintenance costs.
The tariff, or basis of charging the consumer, is designed to recover the above costs as fairly as possible, various types of tariff being available to meet the requirements of various types of consumer.

11.102 A *flat rate* tariff consists simply of a charge of so much per unit (kW h). This type of tariff favours the consumer who uses occasional small amounts of electricity but may prove expensive for the larger consumer. A variation of this tariff is the *block tariff* in which the consumer is charged a high rate for the first "block" of units used during a quarter and a lower rate for any additional units. The result of this tariff is to reduce the cost to the large consumer and to share the fixed costs more fairly between the consumers.

11.103 *Two part tariffs* consist of a fixed charge plus a unit charge, and so can be more closely related to the costs of supply. For domestic consumers the fixed charge is often based on the total floor area of the premises. For certain consumers, e.g. shops, hotels, farms, etc., the fixed charge may be based on the amount of load installed. Industrial consumers often have a charge based on their *maximum demand* in kVA in place of the fixed charge, as a tariff of this type encourages the installation of power factor correction equipment. The maximum demand is usually determined by using a meter which actually measures the maximum number of kVA h used in any half hour period. This type of meter has two pointers; one pointer is reset to zero at half hour intervals and rises to a reading depending on the kVA h used in the following half hour period. This pointer pushes up the second pointer which remains at the highest point reached. Thus at the end of the quarter the second pointer indicates the highest consumption which occurred during any half hour period of the quarter concerned.

11.104 At certain times of the day power stations and mains are not fully used. Electricity suppliers encourage the use of electricity during these "off-peak" periods by offering special "off-peak" tariffs which allow electricity to be supplied for purposes such as storage heating, electric vehicle battery charging, farm crop drying etc., at a low price per unit. When apparatus is supplied on an off-peak tariff basis it may only be used during certain specified periods, the main off-peak period being from 11 p.m. to 7 a.m. B.S.T. The apparatus is controlled by a time switch so that it may only be switched on during the specified period.

Example 11.5

An installation comprises the following: Two 3 kW electric fires, eight 75 W lamps and ten 100 W lamps. It is estimated that the average use of both fires will be 2 hours daily for 365 days, and that the average use of the lamps will be 25 hours weekly for 52 weeks.

Calculate (a) the annual cost with each of the tariffs given below:
(b) the average price per unit with tariff (ii).

Tariff (i): Flat rate. 9p per kW h for heating and 2·5p per kW h for lighting.

Tariff (ii): All-in rate. £4·50 per annum plus 0·5p per kW h for all purposes.

Solution
(a) Lighting:
$$\text{Total power} = (8 \times 75\,\text{W}) + (10 \times 100\,\text{W})$$
$$= 1600\,\text{W or } 1 \cdot 6\,\text{kW}$$
$$\text{kW h consumed per annum} = 1 \cdot 6 \times 25 \times 52$$
$$= 2080\,\text{kW h}$$

Heating:
$$\text{Total power} = 2 \times 3\,\text{kW}$$
$$= 6\,\text{kW}$$
$$\text{kW h consumed per annum} = 6 \times 2 \times 365$$
$$= 4380\,\text{kWh}$$

Tariff (i):
$$\text{Lighting charge} = 2080 \times 2 \cdot 5\text{p}$$
$$= \pounds 52$$
$$\text{Heating charge} = 4380 \times 1\text{p}$$
$$= \pounds 43 \cdot 80$$
$$\text{Total cost} = \pounds 52 + \pounds 43 \cdot 80$$
$$= \pounds 95 \cdot 80$$

Tariff (ii):
$$\text{Total consumption} = 2080 + 4380$$
$$= 6460\,\text{kW h}$$
$$\text{kW h charge} = 6460 \times 0 \cdot 5\text{p}$$
$$= \pounds 32 \cdot 30$$
$$\text{Total cost} = \pounds 32 \cdot 30 + \pounds 4 \cdot 50$$
$$= \pounds 36 \cdot 80$$

(b) Average price per unit $= \dfrac{\pounds 36 \cdot 80}{6460}$
$$= 0 \cdot 57\text{p}$$

11.11 The Cathode Ray Tube

11.111 The cathode ray tube is an electronic device which is used for such purposes as displaying the waveforms of electrical quantities, for displaying television pictures, and for many other purposes. Figure 11.18 shows in simplified form the construction of a cathode ray tube as used in a cathode ray oscilloscope, which is an instrument for displaying electrical waveforms. The electron gun produces a focussed beam of electrons which travel down the evacuated glass tube and strike a fluorescent screen, which is a coating on the inside of the end face of the tube. The electrons excite the screen coating, causing a small spot of light to be produced at the point where they strike the screen.

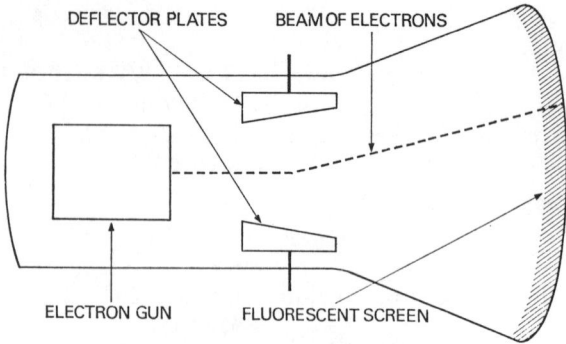

FIGURE 11.18   Cathode Ray Tube.

11.112   The electron gun has much in common with the diode valve. It has a cathode which when heated emits electrons, and an anode which is held at a high positive potential (often several thousand volts) in order to attract the electrons. The anode has a hole in its centre so that most of the electrons pass through it and so travel on to strike the screen. In addition there are often two further anodes. Adjustment of the voltage supplying the central anode can be used to control the focus of the electron beam, and so obtain the smallest possible light spot on the screen. Finally a special control electrode is fitted around the cathode; a negative voltage applied to this electrode reduces the number of electrons leaving the cathode, and this reduces the brightness of the spot of light on the screen, so providing a brilliance control. The arrangement of these electrodes is shown in Fig. 11.19.

11.113   The usefulness of the cathode ray tube stems from the fact that it is very easy to deflect the electron beam and so move the spot of light on the screen. In the tube shown by Fig. 11.18 this is achieved by means of deflector plates. As the electron beam is composed of electrons it is negatively charged, and so it is attracted by a positively charged plate

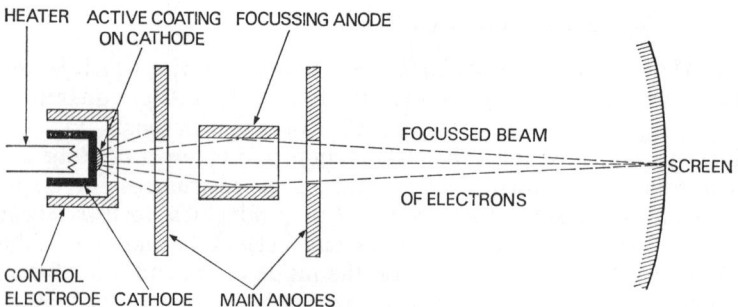

FIGURE 11.19   Construction of Electron Gun.

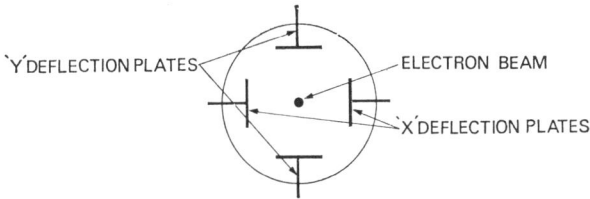

FIGURE 11.20   Arrangement of Deflector Plates in Cathode Ray Tube.

and repelled by a negatively charged plate. This type of deflection is known as *electrostatic deflection*. The tube is usually fitted with two pairs of plates as shown in Fig. 11.20. Voltages applied to the "Y" plates cause the electron beam to be deflected vertically up or down, while voltages applied to the "X" plates cause the beam to be deflected horizontally to right or left. If an alternating voltage is applied to the "Y" plates, the spot of light on the screen is deflected first up and then down in sympathy with the voltage. The result is to produce a vertical line on the screen as shown in Fig. 11.21(a), the length of this line depending on the peak to

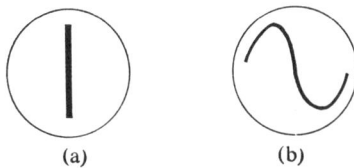

(a)   (b)

FIGURE 11.21   Appearance of Tube Screen.
(a) With an a.c. voltage applied to the "Y" deflector plates only.
(b) With an a.c. voltage applied to the "Y" deflector plates, and a time-base voltage applied to the "X" deflector plates.

peak value of the a.c. voltage. In order to obtain more information about the voltage being examined, the spot of light can be moved horizontally across the screen at a constant speed by applying a suitable voltage to the "X" deflector plates. This causes the spot to trace out a graph of voltage against time. For example if the voltage tested had a frequency of 50 Hz, and the spot took exactly $\frac{1}{50}$ second to cross the screen, then it would trace out one cycle of the a.c. waveform as shown in Fig. 11.21(b). The voltage which must be applied to the "X" plates in order to give the necessary horizontal movement is generated by an electronic circuit called a "time-base" circuit.

11.114   The cathode ray oscilloscope is an instrument which incorporates a cathode ray tube and the necessary time-base circuits. It is usually provided with an amplifier so that small voltages can be magnified if necessary to give a trace of suitable height, the amplifier being fitted with a control which is used to adjust the height of the trace on the screen.

The time-base circuit is fitted with a speed control, so that waveforms of different frequencies can be examined; it is also arranged so that after tracing one picture, the spot flies back to the beginning and starts again. The starting of the spot can be *synchronised* so that it always starts from the same point in the cycle of the waveform being examined; in this way a steady picture of the waveform can be obtained. As the electron beam is virtually weightless it can be deflected very rapidly, and so is capable of accurately following waveforms with frequencies up to several million hertz.

11.115 An alternative method of deflecting an electron beam is to use a magnetic field. *Electro-magnetic* deflection is usually achieved by mounting deflection coils near to the tube. As the beam of moving electrons constitutes an electric current it will experience a force when in a magnetic field, this force acting to move the beam at right angles to its own direction of motion and to the lines of force of the magnetic field. This is the same as the force on a conductor effect described in Section 7.2. Magnetic deflection is not often used in oscilloscopes as the inductance of the coils causes problems at high frequencies, but it is often used in television receivers as the use of magnetic deflection leads to shorter tubes and brighter pictures.

EXERCISE 11

1. Describe, with sketches, a switch board type moving-iron ammeter. State whether it can be used with d.c., or a.c., or with both. Explain how damping, balancing and control are obtained.   C.G.L.I.

2. Explain what is meant by *control* and *damping* as applied to an electrical instrument. Describe how this is effected in the case of a moving-coil and a moving-iron instrument. Illustrate your answer with sketches.   U.L.C.I.

3. (a) Describe, with the aid of a sketch, a *moving-iron* repulsion-type instrument. Can this instrument be used to measure alternating current as well as direct current and if so, why?
   (b) A *moving-coil* ammeter of resistance 5 Ω, requires 15 mA to give full scale deflection. Determine the size of the shunt resistance required in order that the instrument may measure 0 to 50 A.   W.J.E.C.

4. Make a clearly labelled diagram or sketch showing the construction of a moving-coil meter. Describe its action, and explain why it has a uniform scale.
   A moving-coil meter of resistance 5 ohms gives full-scale deflection at a potential difference of 0·075 V. Show how you could adapt the instrument to read up to 240 V. Calculate the value of the extra resistance needed.   C.G.L.I.

5. Draw a sketch and describe the internal construction and action of an instrument for the measurement of the insulation resistance of an installation.
   Explain why the instrument must have a high voltage generator.   E.M.E.U.

6. Describe and explain with the aid of diagrams, how the resistance of a coil of wire may be measured by *each* of the following methods:
   (a) slide-wire bridge;
   (b) ammeter and voltmeter.
   Explain what correction you would make to allow for the instrument resistances.   C.G.L.I.

7. (a) Illustrate by sketches connections of the following instruments when used to measure the input to a single-phase motor:

   (i) Voltmeter,
   (ii) Ammeter,
   (iii) Wattmeter.

   (b) Discuss the use of "moving-coil" and "moving-iron" instruments on a.c. and d.c. circuits. C.G.L.I.

8. When measuring a resistor using the voltmeter and ammeter method, the voltmeter, which had a resistance of 200 Ω was connected directly across the resistor being tested. The ammeter reading was 0·6 A when the voltmeter reading was 20 V. Calculate the true current flowing through the resistor under test and hence find its value in ohms.

9. A small industrial consumer has the following load:

   5 kW of lighting,
   10 kW of heating,
   7·5 kW of small electric motors having an overall efficiency of 74·6%.

   If his tariff is £20 per quarter and 0·6p per kW h for each kW h consumed, find his annual bill for 300 days in the year and 8 hours in the day.

10. Alternative tariffs are offered to a consumer as follows:
    (a) All-in rate: 25p per quarter plus 0·5p per kW h.
    (b) Flat rate: Lighting, 2p per kW h, Heating, 0·6p per kW h.
    The installation consists of the following apparatus: Five 100 watt lamps, six 75 watt lamps, two 2 kW electric fires. The estimated annual use of the lamps is 12 hours weekly for 50 weeks, and the fires will be used for 15 hours weekly for 50 weeks. Calculate the annual cost on each tariff, and thus show which will be the cheaper. C.G.L.I.

11. Why do the Supply Authorities offer a cheaper tariff to a consumer who requires power during the *off peak* periods?
    An electricity consumer installation consists of the following:

    One 2 kW radiator,
    Four 100 watt lamps,
    Two 200 watt lamps.

    If the radiator is in use 500 hours per year, and the lamps are in use for 600 hours per year, calculate the cost of energy, if the charges are in accordance with the following tariff:

    3p per kW h for the first 60 kW h,
    0·5p per kW h for the next 1200 kW h,
    0·4p per kW h for subsequent kW h.

    State the average cost per kW h.

12. The following apparatus is installed in a small house:

    Two 3 kW radiators.
    Eight 75 W lamps.
    Ten 100 W lamps.
    Four 40 W lamps.

It is expected that the average use of each radiator will be 1·5 hours daily for 365 days, and half of the different types of lamps will be in use for 25 hours a week for 52 weeks.

The tariffs offered are:

(a) Flat rate: Heating, 0·5p per kW h.
    Lighting, 2p per kW h.
(b) All-in rate: 12·5p per quarter, plus 0·4p per kW h.

Calculate the estimated annual cost for each tariff, thus showing which will be the more advantageous to the consumer.

Solutions to Numerical Exercises

Chapter 1 (Page 17)

5.

E	6 V	15 V	60 V	240 V	16 V	150 V	50 V
I	3 A	3 A	5 A	2·4 A	8 A	0·5 A	2 A
R	2 Ω	5 Ω	12 Ω	100 Ω	2 Ω	300 Ω	25 Ω
P	18 W	45 W	300 W	576 W	128 W	75 W	100 W

7. 4·17 A, 57·6 Ω, 694 W.   8. £2·84.
9. (a) 300 A h. (b) 1 080 000 C, 9 kW h.
11. 20 h.

Chapter 2 (Page 35)

1. (a) 60 Ω, 80 Ω, 100 Ω. (b) 120 Ω.
    (c) 13·3 Ω, 15 Ω, 24 Ω. (d) 10·9 Ω.
2. 1920 W, 960 W, 480 W.
3. (a) 18·4 A, 14·8 A, 3·69 A. (b) 320 W.
4. (a) 5 A. (b) 2 A, 1·8 A, 1·2 A, 2·5 A, 2·5 A, 22·5 W, 36 V.
5. 6 Ω.
7. 11·7 Ω.
8. (a) 10 A. (b) 1·67 Ω.
9. (a) 5 V from B to D. (b) 18 Ω. (c) 16·7 A.
10. (a) 12·5 Ω. (b) 240 V. (c) 6 A, 12 A, 18 A.
11. (a) (i) 1·12 A, (ii) 8·06 V.
    (b) (i) 0·193 A, (ii) 1·39 V.
    (c) (i) 0·549 A, (ii) 3·95 V.
12. 0·025 Ω.

Chapter 3 (Page 53)

1. 625 MΩ, 4·4 Ω.   2. 0·138 Ω.
3. 60 000 MΩ, 0·27 Ω.   4. 9·52 A.
5. (a) 200 A. (b) 255 V. (c) 1 kW.
6. 232·7 V, 230·5 V.
7. 504 V, 522 V, 25·2 kW, 47 kW.
8. By calculation 82 mm²; nearest available size is 95 mm².
9. (a) 250 V, 250 V. (b) 236·2 V, 230·7 V. (c) 231·2 V, 220·2 V.
10. 30·6°C.   11. 64·2°C.   12. 3·26 A.

ELECTRICAL PRINCIPLES

Chapter 4 (Page 91)

1. (b) 39·2 kW (c) 95·6 A
3. 10·37 kW, 12·96 kW
5. 1·18 MJ
7. 2·56 A
9. 10 N

2. 147 W, 163 W
4. 54·5%
6. 79·4%
8. 67·3%
10. V.R. = 6, M.A. = 4, Efficiency = 66·7%

11. (i) 360 rev/min (ii) 377 W
13. 8·1 N, at 128° from *OA*.

12. 5 N, 36° 54'.

Chapter 5 (Page 102)

1. 5·02 A
3. (a) 16 800 000 J or 16·8 MJ (b) 4 kW (c) 2·7 p
4. 2 h 40 min.
5. 86%.
6. (a) 0·23p (b) 8·23 A
7. (b) 81 620 J
8. (b) 2 100 000 J or 2·1 MJ
9. 5·25 min.
10. 80%.
11. 61 min 26 s.

Chapter 6 (Page 126)

1. 0·833 Wb/m², 200 A t/m.
5. 1·99 mWb.
7. 4730 A t, 7090 A t.
9. 4·21 mWb.

4. 4·61 mWb.
6. 3·96 A, 568.
8. 1·84 mWb.

10. (a) 4 μF, (b) 44 μF, (c) 6·55 μF.
11. 20 μC, 0·2 V, 2 mA.
12. 2 μF, 1·33 μF.

Chapter 7 (Page 160)

1. 19·2 V.
3. 12 N.
7. 253 V.

2. 24 N.
4. 0·096 V.

9. (b) (i) 154 A, 4 A. (ii) 146 A, 4 A.
10. 300 W.   12. 240 V; 216 V, 68 A.

Chapter 8 (Page 192)

1. (a) 191 V; 212 V. (b) 3330 W.
3. (a) 62·8 Ω. (b) 80·3 Ω. (c) 2·49 A. (d) 51·5° lagging.
4. 62·8 Ω, 80·3 Ω, 3·11 A.
5. (a) 60 Ω. (b) 56·6 Ω. (c) 0·178 H. (d) 70·5° lagging.
6. 12·6 Ω, 14·9 Ω, 16·1 A, 57·5° lagging.
7. (b) (i) 8 Ω, (ii) 30 A, (iii) 19·4 A.
8. (a) 30 Ω. (b) 0·127 H. (c) 50 Ω.
9. (a) 17·8 Ω. (b) 0·03 H. (c) 13·5 A. (d) 0·844 lagging.
10. (a) 25·9 A. (b) 5370 W. (c) 0·86 lagging.
11. 28·9 A.
12. $I = \dfrac{E}{\sqrt{R^2 + \left(2\pi fL - \dfrac{1}{2\pi fc}\right)^2}}$, 0·0261 H.

272

Chapter 9 (Page 214)

1. 23·6 A, 16° lagging.
2. 7·15 A, 9° lagging.
3. 28 A.
4. (a) 140 A. (b) 107 A. (c) 140 A.
5. (a) 7·02 kW. (b) 9·36 kV A; 342 μF.
6. 2·5 A, 1·57 A, 0·796 lagging.
7. 12 kW, 15·2 kV A, 0·79 lagging.
8. 203 μF, 10·5 A.
9. 6 kW, 12 kV A, 433 μF.
10. (i) 42·3 A. (ii) 36·6 A. (iii) 42·3 A.
11. 220 V, 380 V.
12. 25 kV A, 34·8 A, 20·1 A.

Chapter 10 (Page 242)

1. 6590 V, 15·2 A.
2. 236 V, 308 A.
3. 1900 turns, 150 turns.
4. 785 turns.
5. 1000 rev/min.
6. 5·33%.
7. 940 rev/min, 1500 rev/min.

Chapter 11 (Page 268)

3. (b) 0·0015 Ω.
4. 15 995 Ω.
8. 0·5 A, 40 Ω.
9. £380·43.
10. (a) £11·35 (b) £20·4.
11. £8·68, 0·586p per kW h.
12. (a) £39·31 (b) £18·22

Index